STUDIES IN GA[illegible] AND MATHEMATIC[illegible]OMICS

Volumes in this Series:

DIFFERENTIAL GAMES AND OTHER GAME-THEORETIC TOPICS IN SOVIET LITERATURE
by Alfred Zauberman

BIDDING AND AUCTIONING FOR PROCUREMENT AND ALLOCATION
edited by Yakov Amihud

GAME THEORY AND POLITICAL SCIENCE
edited by Peter C. Ordeshook

ECONOMICS AND THE COMPETITIVE PROCESS
by James H. Case

GAME THEORY FOR THE SOCIAL SCIENCES
by Hervé Moulin

ABOUT THE SERIES

Game theory, since its creation in 1944 by John von Neumann and Oskar Morgenstern, has been applied to a wide variety of social phenomena by scholars in economics, political science, sociology, philosophy and even biology. This series attempts to bring to the academic community a set of books dedicated to the belief that game theory can be a major tool in applied science. It publishes original monographs, textbooks and conference volumes which present work that is both of high technical quality and pertinent to the world we live in today.

GAME THEORY FOR THE SOCIAL SCIENCES

by

Hervé Moulin

NEW YORK UNIVERSITY PRESS
NEW YORK and LONDON
1982

Library of Congress Cataloging in Publication Data

Moulin, Hervé, 1950–
Game theory for the social sciences.

(Studies in game theory and mathematical economics)
Translation of: Theorie des jeux pour
l'economie et la politique.
Bibliography: p.
Includes index.
1. Social sciences—Mathematical models.
2. Game theory. I. Title. II. Series.
H61.25.M6813 300'.1'5193 82-3460
ISBN 0-8147-5386-8 AACR2
0-8147-5387-6 pbk.

The medallion on the cover of this series was designed by the French contemporary artist Georges Mathieu as one of a set of medals struck in Paris by the Musée de la Monnaie in 1971. Eighteen medals were created by Mathieu to "commemorate 18 stages in the development of western consciousness." The Edict of Milan in 313 was the first, Game Theory, 1944, was number seventeen.

Manufactured in the United States of America

FOREWORD

Game theory is among the most important mathematical tools used in the social sciences today. This book intends to meet two main goals:

i) to offer the non-specialist a self-contained exposition of the essential concepts of strategic games. The definition and results are entirely rigorous from a mathematical standpoint; however more space is devoted to describe and compare the various equilibrium notions than to explore methods for computing them. Technicalities are mostly discarded. Over 75 exercises and problems, some of them difficult, are offered to the motivated reader.

ii) to justify the usefulness of game theoretical concepts in economic and political theory. For that purpose 36 examples have been carefully chosen for their illustrative power. Their inspiration is mainly micro-economic, with applications to imperfect competition, public goods, voting models, division methods, and so on. In many of our problems, the reader is asked to develop further interpretative arguments from the game model. For these some familiarity with economic reasoning will be helpful.

Several important topics in current game-theoretical research are not considered here. The main lacunae are the games of incomplete information, for I feel that the results there lack the maturity and simplicity required for a text book presentation. The next obvious omission are the value approaches to games in characteristic form, that pertain more in my opinion to the non-strategic methodology of social choice (see Introduction).

ACKNOWLEDGEMENTS

The lecture notes of a graduate course on game theory that I gave at the Ecole Nationale de la Statistique et de l'Administration Economique are the basis of this book. I am especially grateful to Paul CHAMPSAUR who warmly invited me to teaching it at the E.N.S.A.E.

My interest in the subject matter of this book was developed at the Centre de Recherches de Mathématiques de la Décision of the Université Paris-Dauphine where I first learned game theory from Jean Pierre AUBIN and Ivar EKELAND. I am also indebted to the Laboratoire d'Econométrie de l'Ecole Polytechnique whose support allowed this manuscript to materialize through the handsome typing of Vilma de SOUZA, Martine VIDONI and Marie-Helène PONROY.

I am also thankful to Andrew SCHOTTER who convinced me to prepare an English version and to the Institute of Mathematical Studies for the Social Sciences, Stanford, whose hospitality gave me an opportunity to complete it.

CONTENTS

GAME THEORY FOR THE SOCIAL SCIENCES

INTRODUCTION

A game is a mathematical object idealizing collective action : several individual agents (the players) influence a situation (outcome of the game) whereas the interest of the players (their utility for the various possible situations) differ. Antagonism of the various utilities raises conflictual behaviour just as identity of all utilities makes the game a pure coordination problem where cooperation is the only rational behaviour. In most games derived from politics and/or economics, the configuration of utilities is neither strict antagonism nor mere identity. The seller and the buyer both agree that their common interest dictates that they reach agreement on exchange, so long as no one is made worse off by the deal ; but they eagerly compete for the choice of a particular price within these limits. Similarly two moderate voters typically agree to destroy the extremist but struggle fiercely to support one of the two prominent middle-of-the road candidates. A moment's reflexion will convince the reader that most social game-like situations generate tendencies to both conflictual and cooperative behaviour. Our claim is that game theory is a useful logical device to explore these mixed motive situations. It does so by displaying a conceptual panoply formalizing a number of behavioural scenarios from the decentralized non-cooperative ones (Chapter I) to the cooperative agreements by mutual threats (Chapter VI). Thus in a given normal form game, cooperative and non cooperative equilibrium concepts will typically coexist. Comparing them is the ultimate point of the game theoretical analysis, presumably a source of rigorous as well as juicy comments on the behavioural incentives generated by the bare normal form structure of the game.

In most social sciences, models abound where strategic considerations are relevant. Economics is the privileged field of development for applied game theory : this was originally suggested by von Neumann and Morgenstern ; yet the well established reputation of the game approach came up only after the Debreu - Scarf theorem settled the cooperative foundations of the competitive equilibrium. Since then, entire subfields of economic theory (like imperfect competition, or the economics of incentives) developed all of them using game theory itself (see the survey paper of Schotter and Schwödiauer [1980]).

Since Farqharson's book ([1969]) strategically oriented analysis of voting processes have renewed several traditional themes of political theory : although the game models used there are still controversial, there is enough evidence that potential contributions of the game view-point are vast. See the survey book of Brams [1975] and the recent literature on implementation (Moulin [1981]) .

Game oriented thinking is now common in sociology : see Boudon [1979], Crozier [1977]. Clearly the search for equilibrium concepts idealizing the spectrum from non-cooperative to cooperative behaviour bears on fundamental issues of sociology. Typically our view of cooperation as non-binding agreements (Part 2) is implicit in Rousseau's paradigm of the evolution from the "liberté naturelle" to the "liberté civile" (see Rousseau [1755]). See also Durkheim's idealist (as opposed to contractualist) approach to cooperation (Durkheim [1893]). However formal game theoretical models are rare or technically very simple in current sociological production. Still the impact of game theory seems already irreversible, if only for pedagogical convenience.

Some epistemological common places.

Did Von Neumann and Morgenstern, when founding game theory, "settled for the first time a scientifically based explanation of the way through from individual behaviour to collective behaviour ?" (Cazelle [1969]). An obvious objection to this optimistic statement works as follows : although the proposed equilibrium concepts sound plausible and give rise to nice interpretations in the examples, no empirical evidence is provided that the players involved in the game you are describing actually follow the rationale of such or such concept. Besides the establishedlist announced here might very well not be exhaustive.

We reply that axiomatic methods in social sciences are needed as a preliminary tatonnement to isolate the fundamental concepts by means of oversimplified models : "the axiomatic language in social sciences signals a rising science, not an advanced one ; this is the only way to abstract from actual experiences". (Granger [1955]). We view game theory as a methodological tool-box used in economic and political theory to develop models. Its increasing influence in these fields is its unique and sufficient justification. As Hayek points out, "social sciences progress by means of concepts" : game theory is indeed a vivid reservoir of flexible concepts each one of them throwing some light on some aspects of social interaction.

The epistemological features common to all social sciences should be in the mind of any newcomer to game theory : the main one is the "universal irruption of value judgements" (Rickert) illustrated here by such notions as Pareto optimality (implicitly taken to be desirable) the search for stable agreements (stability being viewed as a condition necessary to social consensus), the respect of individual strategic freedom (that should be preserved as much as possible, see Part 2), and so on.

As a corollary, no statement, whatever its level of generality, aims to naked objective validity ; it rather tries to raise up subjective approval by the collectivity (i.e. its however small, intended audience). Henceforth Pygmalion's complex, the traditional sin of hard physical sciences (i.e. tendency of the scientist to reshape reality according to his model, not the other way around) looses its devilish connotation in social sciences. Pygmalion is afterall just an unconvincing theoretician : " With regard to human affairs, things are as their actors think they are ". (Hayek, quoted by Freund [1967]).

To put it differently, the social scientist cannot invoke anymore a role of neutral observer : " We explain natural phenomena, we understand the psychic ones ". (Dilthey, quoted by Freund [1967]) .

The unavoidable "empathie" (Boudon) of the observer and the observed materializes in a "verständnis" phase of the analysis where the motivations of the actors are internalized by the scientist. Next the output is an "ideal-type" (Weber) namely a specific model describing convincingly a limited phenomenon with no reference to an underlying general exhaustive theory. For instance we wish to consider each among the dozen equilibrium concepts presented here as a distinct ideal-type illustrating a particular strategic and/or informational aspect of collective action.

An overview of the book.

Part 1 analyzes the non-cooperative behaviour of individual players. The very data of a normal form game allocates the decision power among the players privately (so that all relevant variables are determined as soon as every player has picked his or her individual strategy). We assume first that no communication is feasible among individuals and describe several logical arguments by which a player can decide which strategy he "should" use : dominant and prudent strategies, when our player ignores of the other's utility (Chapter I), sophisticated equilibrium strategies, when the utilities of all players are public knowledge (Chapter II). These strategies can all be computed by each player independently of the other's behaviour, in a purely atemporal world. However appealing these concepts are, their mathematical properties are bad (dominant and sophisticated equilibrium strategies typically fail to exist). There comes the "second generation" of non-cooperative equilibrium concept, namely Nash's, relying on some form of communication among players, if only through their mutually observed behaviour (Chapter III). Several scenarios justifying the Nash'stability property are on the floor, all requiring a dynamic interpretation (perhaps with two periods only). Contrary to the first generation concepts, the Nash equilibrium outcomes share good mathematical properties. If the players use randomized independent strategies, existence of such outcomes is generally guaranteed (Chapter IV).

The essential motivation from Part 1 to Part 2 (cooperative players) is that individual strategic freedom usually hurts the common interest: a non-cooperative equilibrium (whether it comes from a decentralized behaviour -Chapters I and II- or merely satisfies the Nash stability property -Chapters III, IV-) might fail to be a Pareto optimum. Free control by each player of his own strategy contradicts collective efficiency if (and only if) at each Pareto optimal outcome a selfish profitable deviation can occur that destroys the likelihood of this outcome as the solution of our game. Prisonners' Dilemma (Example 1, Chapter I) is the well-known ideal-type of this contradiction. It is the simplest example of the stability / efficiency dilemma, that we view as the main incentive to cooperation. Much less transparent are the very instruments of cooperation.

To achieve outcomes Pareto superior to the non-cooperative equilibrium, we assume that our players enter non-binding agreements, that is to say behavioural scenarios that do not deprive them of the sovereign right of playing any available strategy. These agreements are then self-enforced by a stability property of the Nash type :"I am not forced to follow our agreement but as long as you guys are faithful to it, I have no incentive to betray". In fact the Nash equilibrium notion itself is the first fundamental example of such a cooperative agreement, and the sequence of concepts introduced along Chapter V develop increasingly subtle variants of it.

"We cooperate because we want to, yet our wilful cooperation raises duties that we did not expect" (Durkheim [1893]). A non-binding agreement fully respects individual freedom of choosing selfishly a strategy. Yet its feasibility relies on some enforced limitation of the communication channels. One possibility is that all kinds of communication are banned after the agreement is reached so that individuals cannot be threatened anymore : this is the context in which the stability concepts of Chapter V emerge. Alternatively, one may require that all deviations from the agreed upon strategies must occur publicly so that mutual threats can be carried out. This leads to a cooperative view of deterrence, to which Chapter VI is devoted.

We do not touch the issue of binding agreements, because this is a non-strategic question raising the entire set of problems to which social choice theory is addressed. Instead we work with games of strategy only. By a binding agreement we mean that the players together agree to play a particular outcome and a supreme authority recognized by all enforces the respect of this promise. When they sign the agreement, all players actually loose the control of their own strategy ; thus the issue of stability of this agreement vanishes because no betrayal, either individual or coalitional, is possible any longer. Since the cooperative process is petrified after the agreement is signed the difficult questions all arise before this signature. In other words since conceivable agreements are not discriminated according to their stability, they are distinguished as more or less "just". Hence the view of cooperation switches from positive (which agreements are stable with respect to such and such information structure ?) to normative (which agreements treat fairly the players, given the existing power structure ?). The underlying postulate is that justice is the ultimate cooperative incentive, in other words belief that the proposed agreement is unfair is a principal motive for cooperation to break down.

If only one statement would remain of this course in strategic game theory, it might be that a given game generates several, not a crowd of, equilibrium concepts depending upon the informational and cooperative patterns. Thus a normal form game, our basic mathematical model, proves to be a rich source of strategic scenarios clarifying the logical connections of such primitive notions as : cooperative versus selfish behaviour, mutual strategic anticipations, tactical and/or cooperative uncertainty, communication by threats and/or promises. Ultimate criticism of the proposed concepts should bear on their relevance to the illustrative examples.

The mathematical treatment of our equilibrium concepts lead to existence results as well as non-existence examples. Yet the non-existence of - say - the α-core in a given normal form game should not be interpreted as the logical impossibility of cooperation by threats within this game. It rather means that the deterring scenario encompassed in the α-core concepts does not describe a consistent coalition behaviour so that other more elaborate stories (typically two step concepts : see Chapter VI, Section 2) are in order. Careful avoidance of apocalyptic interpretation is the humility of the model-maker.

REFERENCES

BRAMS, S. 1975. Game theory and politics. London : Glencoe Free Press Collier Macmillan Pub.

CAZELLE, P. 1969."Y a-t-il une science des décisions ?" La Nouvelle Critique n° 24 : 21-27 et n° 25 : 69-75.

CROZIER, M. and F. FRIEDBERG. 1977. L'acteur et le système. Paris : ed. Seuil.

DURKHEIM, E. 1893. De la division du travail social. Paris.

FARQHARSON, R. 1969. Theory of voting. New Haven : Yale University Press.

FREUND, J. 1967. Les théories des sciences humaines. Paris : Presses Universitaires de France.

GRANGER, G. 1955. Méthodologie économique. Paris : Presses Universitaires de France.

MOULIN, H. 1981. "The strategy of social choice". Cahier du Laboratoire d'Econométrie, Ecole Polytechnique, n°A229, Paris. forthcoming in North-Holland Publishing Co.

ROUSSEAU, J.J. 1755. Discours sur l'origine des inégalités.

SCHOTTER, A. and G. SCHWÖDIAUER. 1980. "Economics and game theory : a survey." Journal of Economic Litterature 18, 2.

PART I : NON-COOPERATIVE PLAYERS

The basic model to analyze non-cooperative behaviour is that of a normal form game.

Definition.

Let N be a fixed, finite society, namely a set of players (agents) with index i.

A N-normal form game is defined by the following data :

- for each player $i \in N$ a strategy set X_i with elements x_i
- for each player $i \in N$ a utility function (or payoff) u_i, namely a mapping from $X_N = \underset{i \in N}{\times} X_i$ into $\mathbb{R}$.

An element $x = (x_i)_{i \in N}$ of X_N is called an outcome of the game.

Agent i freely selects a strategy $x_i \in X_i$. Once every agent has exercised his or her strategic power, an outcome x is reached and agent i's utility level is then $u_i(x)$. Notice that no "chance move" nor any "Nature player" exist in the above model.

CHAPTER I. NON-COOPERATIVE DECENTRALIZED BEHAVIOUR.

Given a N-normal form game (X_i , u_i ; $i \in N$) we assume that the players'behaviour is decentralized : each player picks a strategy independently and ignores other players'strategic choices. The players do not communicate. Outcomes are not discriminated for historical reasons (like records of previous plays, or an existing initial position). On the contrary all strategies a priori are equally likely to occur so that discrimination among them must follow from endogeneous rationality arguments.

Throughout Chapter I we assume that a player is aware of his or her own utility function but not necessarily of the other agents'utilities. In this framework we explore successively two consistent decentralized behavioural scenarios : first the elimination of dominated strategies, next the prudent (maximin) behaviour.

I. DOMINATING AND UNDOMINATED STRATEGIES

Definition 1.

Given a N-normal form game (X_i , u_i ; $i \in N$) we say that for player i, strategy x_i dominates strategy y_i if we have :

$$\forall\, x_{\hat{i}} \in X_{\hat{i}} \qquad u_i\,(y_i\,,\, x_{\hat{i}}) \leq u_i\,(x_i\,,\, x_{\hat{i}})$$
$$\exists\, x_{\hat{i}} \in X_{\hat{i}} \qquad u_i\,(y_i\,,\, x_{\hat{i}}) < u_i\,(x_i\,,\, x_{\hat{i}})$$

where $X_{\hat{i}} = \underset{j \in N \setminus \{i\}}{\times} X_j$ and $(x_i\,,\, x_{\hat{i}}) \in X_N$

We denote by $\mathcal{D}_i(u_i)$ the set of agent i's <u>non</u> dominated strategies :

$y_i \in \mathcal{D}_i(u_i) \Leftrightarrow \nexists\, x_i \in X_i$: x_i dominates y_i.

Player i's strategy x_i dominates y_i if no matter how the "rest of the world" $N \setminus \{i\}$ behaves it never pays more to play y_i than x_i and for some feasible strategical choice of $N \setminus \{i\}$ it pays strictly more to play x_i than y_i. This suggests that player i should pick his strategy within $\mathcal{D}_i(u_i)$.

Notice that to compute $\mathcal{D}_i(u_i)$ player i needs only to know the strategy sets X_j of his fellow players, not their utility functions.

<u>Definition 2.</u>

In the N-normal form game $(X_i\,,\, u_i\,;\, i \in N)$ we say that x_i is a <u>dominating strategy</u> of player i if we have $\forall\, y_i \in X_i \;\; \forall\, x_{\hat{i}} \in X_{\hat{i}} : u_i\,(y_i\,,\, x_{\hat{i}}) \leq u_i\,(x_i\,,\, x_{\hat{i}})$.

We denote by $D_i(u_i)$ the set of agent i's dominating strategies.

We say that outcome $x = (x_i)_{i \in N}$ is a <u>dominating strategy equilibrium</u> if x_i is a dominating strategy of player i for all i.

Under mild topological assumptions, non dominated strategies exist : postulating that a non cooperative player eliminates his dominated strategies does not lead to a logical contradiction.

<u>Lemma 1</u>,

Suppose that X_i is compact and u_i is continuous for all $i \in N$. Then the set $\mathcal{D}_i(u_i)$ of player i's undominated strategies is non empty.

<u>Proof</u>.

For all j, pick a probability distribution μ_j on X_j such that every non empty open subset of X_j has positive measure. Fix i and consider the function ψ_i on X_i :

$$\psi_i(x_i) = \int_{X_{\hat{i}}} u_i\ (x_i\ ,\ x_{\hat{i}})\ d\ \mu_{\hat{i}}(x_{\hat{i}})$$

where $\mu_{\hat{i}} = \underset{j \neq i}{\otimes} \mu_j$ is the (tensor) product of μ_j , $j \neq i$.

Since u_i is continuous, ψ_i is continuous as well hence we can choose $x_i^{\star}$ maximizing ψ_i over X_i. We claim that

x_i^* is not dominated. For if x_i dominates x_i^* there exists, by continuity of u_i, an open subset $O_{\hat{\imath}}$ of $X_{\hat{\imath}}$ such that :

$$\forall\, x_{\hat{\imath}} \in O_{\hat{\imath}} \quad u_i\,(x_i^*\,,\, x_{\hat{\imath}}) < u_i\,(x_i\,,\, x_{\hat{\imath}})$$

By our choice of μ_j , $j \neq i$, this implies :

$$\int_{O_{\hat{\imath}}} u_i\,(x_i^*\,,\, x_{\hat{\imath}})\, d\,\mu_{\hat{\imath}}(x_{\hat{\imath}}) < \int_{O_{\hat{\imath}}} u_i\,(x_i\,,\, x_{\hat{\imath}})\, d\,\mu_{\hat{\imath}}(x_{\hat{\imath}})$$

As the inequality $u_i\,(x_i^*\,,\, x_{\hat{\imath}}) \leqslant u_i\,(x_i\,,\, x_{\hat{\imath}})$ holds everywhere on $X_{\hat{\imath}}$ we get :

$$\int_{O_{\hat{\imath}}^c} u_i\,(x_i^*\,,\, x_{\hat{\imath}})\, d\,\mu_{\hat{\imath}}(x_{\hat{\imath}}) \leqslant \int_{O_{\hat{\imath}}^c} u_i\,(x_i\,,\, x_{\hat{\imath}})\, d\,\mu_{\hat{\imath}}(x_{\hat{\imath}})$$

Adding up the above two inequalities yields $\psi_i(x_i^*) < \psi_i(x_i)$ a contradiction.

■

In contrast with the above result, we observe that dominating strategies may fail to exist even for games that are topologically trivial (e.g. finite games). Namely a strategy x_i is dominating if it is a common solution to the maximization problems :

$$\max_{x_i \in X_i} u_i\ (x_i\ ,\ x_{\hat{i}})$$

for all values of the parameter $x_{\hat{i}} \in X_{\hat{i}}$.

Say that strategies x_i and y_i of player i are <u>equivalent</u> if they are not distinguishable in player i's opinion :

$$\forall\ x_{\hat{i}} \in X_{\hat{i}} : u_i\ (x_i\ ,\ x_{\hat{i}}) = u_i\ (y_i\ ,\ x_{\hat{i}})$$

<u>Lemma 2</u>.

Suppose that in the N-normal form game $(X_j\ ,\ u_j\ ;\ j \in N)$ player i's non dominated strategies exist : $\mathcal{D}_i(u_i) \neq \emptyset$. Then the three following properties are equivalent :

i) A dominating strategy of player i exists : $D_i(u_i) \neq \emptyset$

ii) All strategies in $\mathcal{D}_i(u_i)$ are equivalent

iii) $D_i(u_i) = \mathcal{D}_i(u_i)$.

The proof of Lemma 2 is left as an (elementary) exercize to the reader.

Lemma 2 says that when a player has at least one dominating strategy, they are all equivalent and coincide with his non dominated strategies. In this case we postulate that a non cooperative agent will use one of these. On the other hand, if player i has no dominating strategy (by far the most frequent case), then his non dominated strategies are not equivalent, therefore his non cooperative behaviour cannot be defined straightforwardly. Additional informational assumptions (about mutual utility functions) are needed to say more : see Section II below and Chapter II.

An equilibrium in dominating strategies is the postulated rational behaviour of non cooperative players.

Actually in Example 2 and 3 below the dominating strategy equilibrium is Pareto dominated (see Definition 3) thus making this equilibrium objectionable by cooperative arguments.

Example 1. Prisonners' Dilemna (Luce and Raiffa [1957]).

Each of the two players is endowed with two strategies A , P where A stands for agressive and P for peaceful. We assume that "peace" (both players peaceful) is better for both players than "war" (both players agressive) but unilateral agression (one player agressive while the other is peaceful) is profitable to the agressor. A typical payoff structure is the following :

		player 2: P_2	player 2: A_2
player 1	P_1	2 / 2	0 / 3
	A_1	3 / 0	1 / 1

Player 1's strategies are the rows of the payoff matrix whereas Player 2's strategies are its columns. An entry of the 2 × 2 matrix is an outcome of the game. There the North-West number is player 1's utility and the South-East number is Player 2's utility. For instance $u_1 (A_1 , P_2) = 3$.

For both players strategy A dominates strategy P. Hence (A_1 , A_2) is the (unique) dominating strategy equilibrium, therefore war is the postulated non cooperative outcome. However (P_1 , P_2) (peace) provides a better utility level to <u>both</u> players. Therefore non cooperative selfish rationality conflicts with collective interest arguments. Collective interests demand sticking to the peaceful strategy ; yet if the players do not communicate (so that bilateral threats of the form : "I shall be peaceful just as long as you are peaceful" cannot emerge -see Chapter VI) war is the likely outcome : decentralization of the strategic choices imply a transparent collective cost.

<u>Definition 3</u>.

In the N-normal form game $(X_i , u_i ; i \in N)$ we say that outcome $x \in X_N$ <u>Pareto dominates</u> outcome $y \in X_N$ if we have :

$$\begin{cases} \forall\, i \in N & u_i(y) \leqslant u_i(x) \\ \exists\, i \in N & u_i(y) < u_i(x) \end{cases}$$

We say that outcome x is a Pareto optimum if it is not Pareto dominated.

In our next example, the dominating strategy equilibrium is, again, Pareto dominated.

Example 2. Auction of an indivisible good.

An object is to be allocated among n agents. Its value to agent i is a_i and the agents are ordered in such a way that :

$$0 \leqslant a_n \leqslant a_{n-1} \leqslant \dots \leqslant a_2 \leqslant a_1$$

In the sealed bid first price auction each agent independently proposes a price -say agent i proposes x_i- and the winner is the highest bidder who pays its announced price. We denote by r the seller's reservation price, that is the object's value to the seller and we assume that $r \leqslant a_n$. Hence the following n-person game :

$$\begin{cases} X_1 = \dots = X_n = [r , +\infty[\\ \text{For any } x \in X_{\{1,\dots,n\}} \text{ we set } w(x) = \{i / \ x_i = \sup_{1 \leqslant j \leqslant n} x_j\} \\ \text{Then : } u_i(x) = a_i - x_i \quad \text{if } \ i = \inf_{j \in w(x)} j \\ \qquad\qquad = 0 \qquad\qquad \text{otherwise.} \end{cases}$$

Note that we assume no income effect : utility functions are linear with respect to money. Note also that ties are broken in favour of the player who values more the object, a convention with negligible effect.

Observe that a_i, the sincere strategy of agent i dominates every strategy x_i such that $a_i \leqslant x_i$. Namely $u_i(x_i\ ,\ x_{\hat{i}}) \leqslant 0 = u_i\ (a_i\ ,\ x_{\hat{i}})$ for all $x_{\hat{i}}$. Thus $\mathcal{D}_i(u_i) = [r\ ,\ a_i]$. However player i has no dominating strategy unless $r = a_i$: it might pay to underestimate the value of the object.

To eliminate dominated strategies, only one's own utility does matter. If agent 1 is also aware of the object's value to the other agents, he can further eliminate the strategies in $]a_2\ ,\ a_1]$. We explore this line of argument in Chapter II : see in particular Exerçize 3. In the current chapter, however, we focus on decentralized behaviour that relies only on one's own utility.

In the <u>sealed bid second price auction</u> or <u>Vickrey auction</u> the highest bidder wins the object but is charged only the second highest price. Formally:

$$\begin{cases} X_1 = \dots = X_n = [r\ ,\ +\infty[\\ \text{For any } x \in X_{\{1,\dots,n\}} \text{ we write } x_{-i} = \sup\limits_{\substack{1\leqslant j\leqslant n \\ j\neq i}} x_j \\ \text{Then : } u_i(x) = a_i - x_{-i} \quad \text{if} \quad i = \inf\limits_{j\in w(x)} j \\ \qquad\qquad\quad = 0 \qquad\qquad \text{otherwise.} \end{cases}$$

There we claim that *the sincere strategy* a_i *is a dominating strategy of player i*. Namely fix an outcome $x \in X_N$ and distinguish two cases :

Case 1 : player i wins the auction at outcome $(a_i , x_{\hat{i}})$

This implies $x_{-i} \leq a_i$. Observe that $u_i(x)$ is $a_i - x_{-i}$ if i still wins the object at x and zero otherwise. Hence

$$u_i(x_i , x_{\hat{i}}) \leq a_i - x_{-i} = u_i (a_i , x_{\hat{i}})$$

Case 2 : player i does not win the auction at $(a_i , x_{\hat{i}})$

Then $a_i \leq x_{-i}$ and $u_i (x_i , x_{\hat{i}})$ is $(a_i - x_{-i})$ or zero. Therefore :

$$u_i (x_i , x_{\hat{i}}) \leq 0 = u_i (a_i , x_{\hat{i}})$$

One checks easily that no other strategy of i is dominating. Thus $D_i(u_i) = \{a_i\}$ and at the equilibrium in dominating strategies player 1 gets the object and pays a_2. This outcome is Pareto dominated by outcome (r, ... , r) : if every agent announces the reservation price agent 1 still gets the object but pays only r !

Surely agents {2 , ... , n} will not be so nice to agent 1 for free so that agent 1 should redistribute part of his extra-profit $a_2 - r$ to the remaining players : this cooperative approach to auction is irrelevant now : see Chapter VI, Section 5.

Example 3. Don't forget the favour.

If some agent has several dominating strategies, they are equivalent to him (Lemma 2), perhaps not to the others. Consider the following two persons game where an agent's strategy influences only the other's utility :

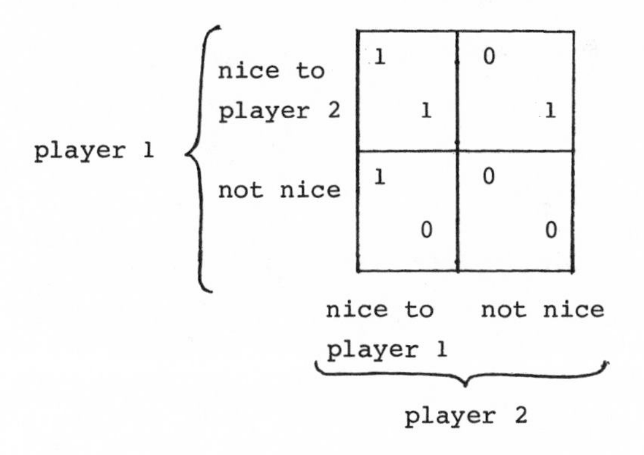

Every outcome is a dominating strategy equilibrium yet only one outcome is Pareto optimal.

Problem 1. Two oligopoly games. (Case [1979]).

1) A price-setting duopoly game :

The two duopolists offer two substitutable goods. If they set the prices p_1 , p_2 the corresponding demands are :

$$d_1 = \left(\frac{p_2}{p_1}\right)^{\alpha_1} \quad \text{units of the good produced by player 1}$$

$$d_2 = \left(\frac{p_1}{p_2}\right)^{\alpha_2} \quad \text{units of the good produced by player 2}$$

We assume a constant return to scale technology for both producers. Thus the following normal form game :

$$u_i\ (p_1\ ,\ p_2) = (p_i - c_i)\ .\ d_i \qquad \text{where } c_i \text{ is constant.}$$

Compute the equilibrium in dominating strategies of this game and comment on the assumed form of the demand functions.

2) A quantity-setting oligopoly :

Suppose the price of a typical satiable good, say mineral water, goes to $c\ e^{-S}$ where S is the total supply. If n costless producers control the quantities $x_1 , \dots , x_n$ of mineral water that they supply we obtain the following game

$$u_i\ (x_1\ ,\ \dots\ ,\ x_n) = c\ x_i\ e^{-(x_1+\dots+x_n)}$$

Answer the same questions as in 1).

Problem 2. *Topological properties of $\mathcal{D}_i$ and D_i .*

For all $i \in N$ let X_i be a compact set and u_i be a continuous function on X_N.

1) Prove that agent i has a unique undominated strategy if and only if he has a unique dominating strategy.

2) Prove by an example that the sets $\mathcal{D}_i(u_i)$ of undominated strategies and the set of Pareto optimal outcomes are not necessarily closed. What about the sets $D_i(u_i)$ of dominating strategies ?

Problem 3. *Strategy proof voting on a line.*
Moulin [1980].

Let $(A , \lesssim)$ be an ordered set of p candidates among which society $N = \{1 , 2 , \ldots , n\}$ must select one. Given any <u>odd</u> integer K we denote by m_K the following mapping from A^K into A :

$$m_K (a_1 , \ldots , a_K) = b \Leftrightarrow \begin{cases} | \{k / a_k \lesssim b\} | \geqslant \frac{K + 1}{2} \\ | \{k / a_k \gtrsim b\} | \geqslant \frac{K + 1}{2} \end{cases}$$

for all $(a_1 , \ldots , a_K) \in A^K$

Thus $m_K (a_1,\ldots, a_K)$ is the median element of $\{a_1,\ldots,a_K\}$ according to the given ordering of A .

We define now a family of voting rules on A where player i's strategy set (or message set) is simply $X_i = A$ and the decision rule takes the form :

$$\begin{cases} \pi(x_1, \dots, x_n) = m_{2n-1}(x_1, \dots, x_n, \alpha_1, \dots, \alpha_{n-1}) \\ \text{where } (x_1, \dots, x_n) \text{ varies over } X_{\{1,\dots,n\}} \text{ and} \\ \alpha_1, \dots, \alpha_{n-1} \text{ are fixed elements of A.} \end{cases}$$

In particular setting $\alpha_1 = \dots = \alpha_{n-1} = \inf_{a \in A} a$ yields $\pi(x_1, \dots, x_n) = \inf_{1 \leq i \leq n} x_i$. Another example is $\alpha_1 = \dots = \alpha_j = \inf_{a \in A} a$ and $\alpha_{j+1} = \dots = \alpha_{n-1} = \sup_{a \in A} a$. Then $\pi_{j+1}(x_1, \dots, x_n)$ is the candidate with rank j+1 when $x_1, \dots, x_n$ are listed by decreasing order according to the given ordering of A.

When agent i is endowed with a utility function u_i over A, a voting rule π generates the n-person game :

$$g = (X_1, \dots, X_n ; u_1 \circ \pi, \dots, u_n \circ \pi)$$

1) Let x_i and y_i be two strategies contiguous with respect to the given ordering of A. Prove the following implication :

$$\{u_i(x_i) > u_i(y_i)\} \Rightarrow \{x_i \text{ dominates } y_i \text{ for player i}\}$$

$$\{u_i(x_i) = u_i(y_i)\} \Rightarrow \{x_i \text{ and } y_i \text{ are equivalent for player i}\}.$$

Deduce that $D_i(u_i)$ is made up only of local maxima of u_i :

$$x_i \in D_i(u_i) \Rightarrow [\forall y_i \in A \{u_i(x_i) < u_i(y_i)\} \Rightarrow \exists z_i \in] x_i, y_i [u_i(z_i) < u_i(x_i) \}$$

2) Prove that player i has a dominating strategy if his utility u_i is single-peaked on $(A, \lesssim)$ which is to say :

$$\exists \alpha, \beta \in A \quad \begin{array}{ll} \text{on }]\leftarrow, \alpha] & u_i \text{ is non decreasing} \\ \text{on } [\alpha, \beta] & u_i \text{ is constant} \\ \text{on } [\beta, \rightarrow[& u_i \text{ is non increasing} \end{array}$$

What is the set $D_i(u_i)$ in that case ?

Prove that a dominating strategy equilibrium of g is Pareto optimal.

3) When the utilities u_i are arbitrary on A, prove that undominated strategies in the game g derived from $\pi_1, \ldots, \pi_n$ are as follows :

- For $\pi_2, \ldots, \pi_{n-1}$ $D_i(u_i) = LM(u_i) = \{x_i \in A / x_i \text{ is a local maxima of } u_i\}$

- For π_1 : $D_i(u_i) = \{x_i \in LM(u_i) / \forall y_i \in LM(u_i) [y_i < x_i] \text{ or } [u_i(y_i) < u_i(x_i)]\}$.

Problem 4. *The Clarke-Groves mechanism for financing a public good.* (Green - Laffont [1979]).

Let A be a set of p outcomes (representing various public projects, candidates,...) among which society N must select one. In contrast with the previous problem we suppose that side-payments (i.e. monetary transfers) are possible and agents are endowed with quasi-linear utilities. Thus if u_i, a real valued function on A, represents agent i's utility, then agent i's utility level is $u_i(a) + t_i$ if $a \in A$ is the choosen decision and t_i the (positive or negative) transfer to agent i.

In the Clarke - Groves mechanism each agent's message is to reveal his utility function. As agent i's information on u_i is private, he can pick any $x_i \in \mathbb{R}^A$ and pretend it to be his utility. Hence $X_i = \mathbb{R}^A$ is player i's strategy set as well. After each agent i has sent a message x_i, an outcome $x \in X_N$ results. Then a decision $a^* = a^*(x) \in A$ is selected such that :

$$\sum_{i \in N} x_i(a^*) = \sup_{b \in A} \{ \sum_{i \in N} x_i(b) \} \qquad (1)$$

And the monetary transfer to agent i is $t_i(x)$:

$$t_i(x) = \sum_{j \in N \setminus \{i\}} x_j(a^*) - \sup_{b \in A} \{ \sum_{j \in N \setminus \{i\}} x_j(b) \}$$

Therefore if $(u_i)_{i \in N}$ is the N-uple of true utility functions the agents face the N-normal form game $g = (X_i \; ; \; \tilde{u}_i \; , i \in N)$, where :

$$\tilde{u}_i(x) = u_i(a^*(x)) + t_i(x)$$

Note that the mapping $x \to a^*(x)$ is any mapping satisfying (1).

1) Observe that $t_i(x)$ is never positive and interpret the cases $t_i(x) = 0$ and $t_i(x) < 0$.

2) Prove that the sincere strategy $x_i = u_i$ is the unique (up to the addition of a constant function) dominating strategy of agent i.

3) Prove that the dominating strategy equilibrium $u = (u_i)_{i \in N}$ is Pareto optimal if and only if $t_i(a^*(u)) = 0$ for all i. Interpret this result.

II. PRUDENT AND OPTIMAL STRATEGIES

If player i's information about other agents' utility functions is zero, he can not discriminate at all among the possible strategies in X_i (we assume that the strategy sets of all agents are publicly known).

One way to discriminate among X_i is to eliminate dominated strategies. Another way is to follow the pessimistic (risk-averse) prediction that the worst will happen:

Definition 4.

In the N-normal form game $(X_i, u_i; i \in N)$ we say that x_i is a prudent strategy of player i if we have :

$$\inf_{x_{\hat{\imath}} \in X_{\hat{\imath}}} u_i(x_i, x_{\hat{\imath}}) = \sup_{y_i \in X_i} \inf_{x_{\hat{\imath}} \in X_{\hat{\imath}}} u_i(y_i, x_{\hat{\imath}})$$

We denote by $P_i(u_i)$ the set of player i's prudent strategies.

Prudent behaviour is feasible and consistent with the elimination of dominated strategies :

Lemma 3.

Suppose that X_i is compact and u_i is continuous for all $i \in N$, then the set $P_i(u_i)$ of player i's prudent strategies is non empty, compact and intersects the set $\mathcal{D}_i(u_i)$ of undominated strategies :

$$P_i(u_i) \cap \mathcal{D}_i(u_i) \neq \phi$$

Proof.

We fix $i \in N$. As u_i is continuous, the function $\theta(y_i) = \inf_{x_{\hat{\imath}} \in X_{\hat{\imath}}} u_i(y_i, x_{\hat{\imath}})$ is uppersemicontinuous on X_i and therefore reaches its maximum over a non empty compact set $P_i(u_i)$. Next consider the N-game $H = (Y_j, u_j, j \in N)$ where

$Y_j = X_j$ for all $j \neq i$ and $Y_i = P_i(u_i)$.

By Lemma 1 player i has at least one undominated strategy x_i in H. Suppose x_i is dominated by y_i in the <u>original</u> game (X_i, u_i) : $[\forall x_{\hat{i}} \in X_{\hat{i}} \;\; u_i(x_i, x_{\hat{i}}) \leqslant u_i(y_i, x_{\hat{i}})] \Rightarrow$ $[\theta(x_i) \leqslant \theta(y_i)]$. This implies $\theta(y_i) = \sup\limits_{z_i \in X_i} \theta(z_i)$ and $y_i \in P_i(u_i)$, contradicting our assumption that x_i is undominated in H. Thus x_i belongs to $P_i(u_i) \cap \mathcal{D}_i(u_i)$.

$\blacksquare$

In Problem 5 below we propose a variant of Definition 4 that always selects a subset of $P_i(u_i) \cap \mathcal{D}_i(u_i)$.

By using any prudent strategy, agent i is guaranteed of the utility level $\alpha_i = \sup\limits_{x_i \in X_i} \inf\limits_{x_{\hat{i}} \in X_{\hat{i}}} u_i(x_i, x_{\hat{i}})$. Call α_i his <u>secure</u> utility level. When the N-uple $(\alpha_i)_{i \in N}$ of secure utility levels does not leave room for Pareto improvements, we claim that prudent strategies can be said optimal as well :

<u>Definition 5</u>,

We will say that the N-normal form game $(X_i, u_i ; i \in N)$ is <u>inessential</u> if there is no outcome $y \in X_N$ such that :

$$\begin{cases} \forall i \in N \quad \sup\limits_{x_i \in X_i} \inf\limits_{x_{\hat{i}} \in X_{\hat{i}}} u_i(x_i, x_{\hat{i}}) = \alpha_i \leqslant u_i(y) \\ \exists i \in N \qquad\qquad\qquad\qquad\qquad\qquad \alpha_i < u_i(y) \end{cases}$$

The intuition behind inessential games is as follows : one unit of an homogeneous cake is to be divided among society N. Suppose player i — by playing well — can guarantee that his final share is at least α_i, whatever is the behaviour of other players : suppose moreover that $\sum_{i \in N} \alpha_i = 1$. Then $(\alpha_i)_{i \in N}$ is the sharing of the cake that must result from the "optimal" behaviour by the players. In our general framework, utility levels are not comparable between players, hence can not be added. Nevertheless for an inessential game, prudent strategies are optimal in the following sense :

<u>Theorem 1.</u>

Let $(X_i, u_i ; i \in N)$ be an inessential game. For all $i \in N$, let x_i be a prudent strategy of i, and let x be the associated outcome. Then :

1) $u_i(x) = \alpha_i \leqslant u_i(x_i, y_{\hat{i}})$ for all $i \in N$ and $y_{\hat{i}} \in X_{\hat{i}}$

2) x is Pareto optimal

3) for all coalitions $S \subset N$ and all strategy S-uples $y_S \in X_S$ the following conditions together are impossible :

$$\begin{cases} \forall i \in S & u_i(x) \leqslant u_i(y_S, x_{S^c}) \\ \exists i \in S & u_i(x) < u_i(y_S, x_{S^c}) \end{cases}$$

Proof.

Since x_i is a prudent strategy of agent i, we have :

$$\alpha_i = \inf_{y_{\hat{i}} \in X_{\hat{i}}} u_i(x_i, y_{\hat{i}}) \leq u_i(x)$$

The above inequality holds for all i : as our game is inessential this implies $\alpha_i = u_i(x)$ for all i, hence property 1.

Property 2 follows from 3 by making $S = N$. To prove 3 we pick S and $y_S \in X_S$ such that :

$$\forall i \in S \quad \alpha_i \leq u_i(y_S, x_{S^c}) \qquad (2)$$

Applying property 1 to $j \in S^c$ we get :

$$\forall j \in S^c \quad \alpha_j \leq u_j(y_S, x_{S^c})$$

Combining these two systems of inequalities yields :

$u_i(y_S, x_{S^c}) = \alpha_i$ for all i, because our game is inessential. Thus no inequality can be strict in (2).

■

Property 1 states that if player i uses an optimal (i.e. prudent) strategy and expects the other players to do the same, he enjoys the utility level α_i ; should some players j, $j \neq i$, fail to use an optimal strategy, this can only be profitable to i.

Property 3 means that neither a single player nor a coalition of players has any incentive to unilaterally deviate from prudent strategies (we implicitly assume that no side payments can take place within a coalition : utilities are not transferable). In terms of Definition 1, Chapter V, we say that a N-uple of prudent strategies is a strong equilibrium.

To interpret further Definition 5, consider a <u>non</u> inessential game $(X_i, u_i ; i \in N)$ and remark that no N-uple x can be called optimal.

Namely two obvious optimality requirements are $\alpha_i \leqslant u_i(x)$ for all i, and x is a Pareto optimum. By Definition 5 these conditions together imply that for some $i \in N$ we have :

$$\sup_{y_i} \inf_{y_{\hat{i}}} u_i(y_i, y_{\hat{i}}) = \alpha_i < u_i(x)$$

In words player i is not guaranteed of his pay-off $u_i(x)$ and can be threatened by the complement coalition $N \setminus \{i\}$.

The main example of inessential games are certain two person zero sum games, to which the next section is devoted.

<u>Exercise 1.</u>

Suppose g is a 2-person game with a Pareto optimal dominating strategy equilibrium. Is g necessarily inessential?

Problem 5. *Lex-prudent strategies.*

Moulin [1981] .

Given an element $z = (z_1, \ldots, z_T)$ in $\mathbb{R}^T$ if we reorder its coordinates in increasing order, the resulting vector is denoted $z^\star$:

$$z^\star = (y_1, \ldots, y_T) \Leftrightarrow \{y_1, \ldots, y_T\} = \{z_1, \ldots, z_T\}$$

$$y_1 \leqslant y_2 \leqslant \ldots \leqslant y_T$$

Next we denote by α the lexicographical ordering of $\mathbb{R}^T$:

$$y \ \alpha \ z \Leftrightarrow \exists\, t_o, 1 \leqslant t_o \leqslant T \begin{cases} \forall\, t < t_o & y_t = z_t \\ y_{t_o} < z_{t_o} \end{cases}$$

Let $(X_i, u_i \, ; \, i \in N)$ be a given N - normal form game such that X_i are all finite. We shall say that a strategy x_i of agent i is <u>lex-prudent</u> if it maximizes with respect to the lexicographical ordering of $\mathbb{R}^{X_{\hat{i}}}$ the mapping v_i defined as follows :

$$v_i(x_i) \doteq (u_i(x_i, x_{\hat{i}}) \, / \, x_{\hat{i}} \in X_{\hat{i}})^\star$$

We denote by $LP_i(u_i)$ the set of agent i' lex-prudent strategies.

1) Interpret the above definition

2) Prove the inclusion : $LP_i(u_i) \subset P_i(u_i) \cap \mathcal{D}_i(u_i)$

3) Prove that lex-prudent strategies generalize dominating strategies in the following sense :

$$\{ D_i(u_i) \neq \phi \} \Rightarrow \{ LP_i(u_i) = D_i(u_i) = \mathcal{D}_i(u_i) \}$$

III. TWO-PERSON ZERO-SUM GAMES

A two person zero sum game takes the form $(X_1, X_2 ; u_1, -u_1)$ i.e. the two players are purely antagonistic. We denote it simply (X_1, X_2, u_1) and interpret u_1 as a (positive or negative) pay-off that player 1 maximizes and player 2 minimizes. Thus the prudent strategies are given as follows :

$$x_1 \in P_1(u_1) \Leftrightarrow \inf_{y_2 \in X_2} u_1(x_1, y_2) = \sup_{y_1 \in X_1} \inf_{y_2 \in X_2} u_1(y_1, y_2)$$

$$x_2 \in P_2(-u_1) \Leftrightarrow \sup_{y_1 \in X_1} u_1(y_1, x_2) = \inf_{y_2 \in X_2} \sup_{y_1 \in X_1} u_1(y_1, y_2)$$

The numbers $\sup_{y_1} \inf_{y_2} u_1$ and $\inf_{y_2} \sup_{y_1} u_1$ are respectively the (minimal) secure gain of player 1 and the (maximal) secure loss of player 2. They are related by the inequality :

$$\sup_{y_1} \inf_{y_2} u_1 \leqslant \inf_{y_2} \sup_{y_1} u_1 \tag{3}$$

To prove (3) we fix $x_1 \in X_1$, $x_2 \in X_2$ and we observe :

$$\phi_1(x_1) = \inf_{y_2} u_1(x_1,y_2) \leqslant u_1(x_1,x_2) \leqslant \sup_{y_1} u_1(y_1,x_2) = \phi_2(x_2)$$

This implies :

$$\sup_{y_1} \phi_1(y_1) \leqslant \inf_{y_2} \phi_2(y_2)$$

Theorem 2.

Let $G = (X_1, X_2 ; u_1)$ be a two person zero sum game. If we have $\sup_{y_1} \inf_{y_2} u_1 = \inf_{y_2} \sup_{y_1} u_2$, we shall call this number the value of G. If (3) is a strict inequality, we shall say that G has no value. If G has a value, then it is inessential. Conversely, suppose X_1, X_2 are compact and u_1 is continuous. Then if G is inessential, it has a value.

Proof.

Suppose G has a value and pick an outcome $(x_1, x_2) \in X_1 \times X_2$ such that

$$\begin{aligned} &\sup_{y_1} \inf_{y_2} u_1 \leqslant u_1(x_1, x_2) \\ &\sup_{y_2} \inf_{y_1} -u_1 \leqslant -u_1(x_1,x_2) \Leftrightarrow u_1(x_1,x_2) \leqslant \inf_{y_2} \sup_{y_1} u_1 \end{aligned} \tag{4}$$

Because (3) is an equality, the above two inequalities actually are equalities and the first statement is proved.

Conversely suppose that (3) is strict. By our topological assumptions and Lemma 3 we can pick a pair $(x_1, x_2) \in P_1(u_1) \times P_2(u_2)$. Then (x_1, x_2) satisfies system (4) with at least one strict inequality so that G cannot be inessential.

■

For inessential zero-sum games, pairs of optimal strategies are characterized as saddle-pairs :

Definition 6.

A saddle pair of the two person zero sum game (X_1, X_2, u_1) is a pair $(x_1, x_2) \in X_1 \times X_2$ such that :

$$\forall (y_1, y_2) \in X_1 \times X_2 \quad u_1(y_1, x_2) \leq u_1(x_1, x_2) \leq u_1(x_1, y_2)$$

We denote by S the (possibly empty) subset of saddle pairs.

Theorem 3.

Let $G = (X_1, X_2, u_1)$ be a two person zero sum game. If G has a value then an outcome is a pair of optimal strategies if and only if it is a saddle-pair : $S = P_1(u_1) \times P_2(u_2)$.

If G has no value then it has no saddle-pair either.

<u>Proof.</u>

Suppose first that G has a value, hence is inessential. From property 3 of Theorem 1 we derive the inclusion $P_1(u_1) \times P_2(u_2) \subset S$. Conversely, we pick a saddle-point (x_1, x_2) and deduce from Definition 6 that we have :

$$\inf_{y_2} \sup_{y_1} \leqslant \sup_{y_1} u_1(y_1, x_2) \leqslant u_1(x_1, x_2) \leqslant \inf_{y_2} u_1(x_1, y_2) \leqslant \sup_{y_1} \inf_{y_2} u_2$$

By (3) these 4 inequalities are in fact equalities, implying that x_i is a prudent strategy of player i (i = 1 , 2).

Theorems 2 and 3 prove that the key strategical feature of zero-sum games is the existence or non-existence of a value. If the game possesses a value, then optimal strategies exist and are computed equivalently in a decentralized way (as prudent strategies) or simultaneously for both players (as saddle- pairs). On the other hand games having no value raise a typical non converging sequence of strategical anticipations. Namely, let (X_1, X_2, u_1) be a zero sum game where

$$\sup_{x_1} \inf_{x_2} u_1 = a < b = \inf_{x_2} \sup_{x_1} u_1$$

Say that player 1 wins by forcing the final pay-off above b whereas player 2 wins by keeping the final pay-off below a.

Suppose player 1 considers strategy x_1^o. By anticipating this choice and using a best reply strategy x_2^1, player 2 wins :

$$u_1 (x_1^o , x_2^1) = \inf_{y_2} u_1 (x_1^o , y_2) \leq a$$

By anticipating player 2's choice x_2^1 and using a best reply strategy x_1^1, player 1 wins :

$$b \leq \sup_{y_1} u_1 (y_1 , x_2^1) = u_1 (x_1^1 , x_2^1)$$

And so on ... In a sequence $(x_1^t , x_2^t)_{t \in \mathbb{N}}$ where x_2^t is a best reply of player 2 to x_1^{t-1} and x_1^t is a best reply of player 1 to x_2^t we have

$$u_1 (x_1^{t-1} , x_2^t) \leq a < b \leq u_1 (x_1^t , x_2^t)$$

This implies that neither of the two sequences $(x_1^t)_{t \in N}$ or $(x_2^t)_{t \in N}$ converges (under the assumption that u_1 is continuous and X_1 , X_2 are compact).

We conclude by an example of zero-sum game with and without a value and a few exercises and problems.

Example 4. Tic-tac-toe,

Both players choose a strategy within tic,tac,toe . Player 1's pay-off is strictly positive if he guesses correctly Player 2's choice, and zero otherwise. The utility u_1 is given by the following 3 x 3 matrix :

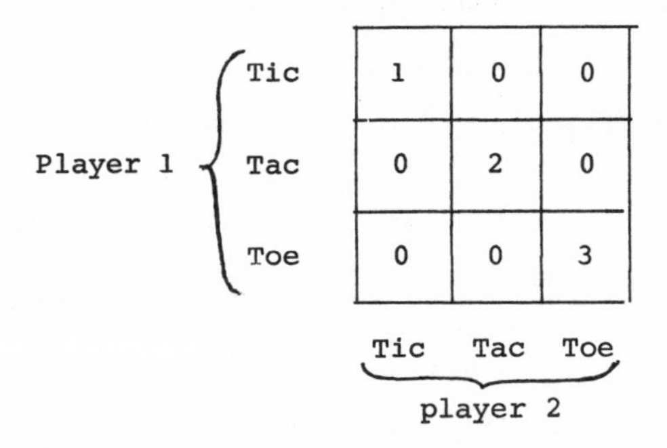

Player 1's secure gain is 0 and anyone of his strategies is prudent (for any x_1 the guaranteed pay-off level is only zero).

Player 2's secure loss is 1 and he has a unique prudent strategy, namely Tic (while playing Tac or Toe he faces the risk of a loss 2 or 3).

In Chapter IV we will use mixed (i.e. random) strategies to extend the above game into one with a value.

Example 5. Gunfight.

Both players have one bullet in their gun and walk toward each other at constant speed. At time $t = 0$ they are far apart and at time $t = 1$ they coincide. The real valued function a_i, defined on $[0, 1]$, measures the skill of player i, $i = 1, 2$. Namely $a_i(t)$ is the probability that player i hits player j if he shoots at time t. We assume that a_i is non decreasing, continuous and that $a_i(0) = 0$, $a_i(1) = 1$. The payoff is + 1 if player 1 hits player 2 before being hurted himself, - 1 in the symmetrical case, and 0 if no one is hurted or both are simultaneously.

The strategy sets are $X_1 = X_2 = [0, 1]$. Strategy x_i for player i means : I will shoot at time $t = x_i$ if the opponent did not shot yet. If he did, and did not reached me, I will shoot safely at time $t = 1$. Thus the normal form of the game is (X_1, X_2, u_1) where :

$$u_1(x_1, x_2) = \begin{cases} 2a_1(x_1) - 1 & \text{if } x_1 < x_2 \\ a_1(x_1) - a_2(x_1) & \text{if } x_1 = x_2 \\ 1 - 2a_2(x_2) & \text{if } x_2 < x_1 \end{cases} \quad (5)$$

For instance assume $x_1 = x_2$. Then the payoff is + 1 with probability $a_1(x_1) \cdot (1 - a_2(x_1))$, i.e. if player 1 hits his opponent and is not hit, and - 1 with probability $a_2(x_1) \cdot (1 - a_1(x_1))$.

Let us compute the prudent strategies of player 1. In view of (5) one checks easily that for all $x_1 \in [0, 1]$

$$\varphi_1(x_1) = \inf_{x_2 \in [0,1]} u_1(x_1, x_2) = \inf \{2 a_1(x_1) - 1 , 1 - 2 a_2(x_1)\}$$

Let I be the following subinterval of [0 , 1] (in general a singleton) :

$$I = \{x_1 \in [0, 1] \ / \ 2 a_1(x_1) - 1 = 1 - 2 a_2(x_1)\}$$

Then φ_1 increases before I, is constant on I, and decreases after I. Therefore $I = P_1(u_1)$ is the set of player 1's prudent strategies. The secure payoff $\alpha_1 = \sup_{x_1} \inf_{x_2} u_1$ is the common value of $2 a_1 - 1$ and $1 - 2 a_2$ on I. A similar computation shows that $I = P_2(u_2)$ as well and the secure payoff of player 2 equals α_1. Therefore it is the value of the game and I is the set of optimal strategies for both players. Each one optimally shoots when $a_1(t) + a_2(t) = 1$.

Exercise 2.

Generalize Theorems 2 and 3 to the class of two person quasi-zero-sum games namely those games (X_1, X_2, u_1, u_2) such that :

every outcome is Pareto optimal, or equivalently :

$$\forall x, y \in X_1 \times X_2 : [u_1(x) \leqslant u_1(y)] \Leftrightarrow [u_2(y) \leqslant u_2(x)]$$

Exercise 3. Silent gunfight.

In this variant of the above game a player's shoot is silent thus his opponent is not aware of it unless he is hit.

1) Prove that the corresponding pay-off function is now :

$$X_1 = X_2 = [0, 1]$$

$$\tilde{u}_1(x_1, x_2) = \begin{cases} a_1(x_1) - a_2(x_2) + a_1(x_1).a_2(x_2) & \text{if } x_1 < x_2 \\ a_1(x_1) - a_2(x_2) & \text{if } x_1 = x_2 \\ a_1(x_1) - a_2(x_2) - a_1(x_1).a_2(x_2) & \text{if } x_2 < x_1 \end{cases}$$

2) Suppose that $a_1 - a_2 + a_1 \cdot a_2$ is an increasing function of t whereas $a_1 - a_2 - a_1 . a_2$ is a decreasing one. Compute the secure utility level of both players and prove that they are different. More precisely, if v denotes the value of the noisy gunfight (Example 4 above) prove that :

$$\sup_{x_1} \inf_{x_2} \tilde{u}_1 < v < \inf_{x_2} \sup_{x_1} \tilde{u}_1$$

Exercise 4.

Let (X_1, X_2, u_1) be a symmetrical two person zero sum game :

$$X_1 = X_2 \;;\; u_1(x_1, x_2) = -u_1(x_2, x_1) \text{ all } x_1, x_2$$

Prove that its value (if any) is zero and the sets of optimal strategies coïncide.

Problem 6. (Moulin [1976]).

For all real numbers a, b, c, d the following two person zero sum game, where each player has four strategies, has a value (the pay-off is that of the row player) :

$$G = \begin{vmatrix} a & b & a & b \\ c & d & d & c \\ a & d & a & \frac{a+b+c+d}{4} \\ c & b & \frac{a+b+c+d}{4} & c \end{vmatrix}$$

Moreover the value of G satisfies :

$$\sup\{\inf(a,b), \inf(c,d)\} \leq \mathrm{val}(G) \leq \inf\{\sup(a,c), \sup(b,d)\}$$

Interpretation ?

Problem 7. (*Shapley*).

Let $G = (X_1, X_2, u_1)$ be a finite two person zero sum game. Suppose that for any subsets $Y_i \subset X_i$, $i = 1, 2$ such that $|Y_i| = 2$, $i = 1, 2$, the game (Y_1, Y_2, u_1) has a value. Prove that G itself has a value.

Problem 8.

Let (X_1, X_2, u_1) be a two person zero-sum game. We denote by $R_1 = X_1^{X_2}$ (resp. $R_2 = X_2^{X_1}$) the set of reply strategies of player 1 (resp. player 2).

1) Prove the following equalities :

$$\sup_{x_1 \in X_1} \inf_{x_2 \in X_2} u_1(x_1, x_2) = \inf_{r_2 \in R_2} \sup_{x_1 \in X_1} u_1(x_1, r_2(x_1))$$

$$\inf_{x_2 \in X_2} \sup_{x_1 \in X_1} u_1(x_1, x_2) = \sup_{r_1 \in R_1} \inf_{x_2 \in X_2} u_1(r_1(x_2), x_2)$$

What are the optimal strategies of the game $(R_1, X_2, \tilde{u}_1)$ (where we set $\tilde{u}_1(r_1, x_2) = u_1(r_1(x_2), x_2)$) ?

2) We suppose that X_1, X_2 are convex and compact and denote by $R_1^c \subset R_1$ (resp. $R_2^c \subset R_2$) the set of continuous reply functions from X_2 into X_1 (resp. from X_1 into X_2).

Prove the following inequalities :

$$\sup_{x_1} \inf_{x_2} u_1 \leqslant \sup_{r_1 \in R_1^c} \inf_{x_2} u_1(r_1(x_2), x_2) \leqslant \inf_{r_2 \in R_2^c} \sup_{x_1} u_1(x_1, r_2(x_1))$$

$$\ldots \leqslant \inf_{x_2} \sup_{x_1} u_1$$

(Use Brouwer's theorem : a continuous mapping from a convex compact set into itself has at least one fixed point).

3) Give an example where $X_1 = X_2 = [0,1]$, u_1 is continuous and moreover :

$$\sup_{x_1} \inf_{x_2} u_1 = \sup_{r_1 \in R_1^c} \inf_{x_2} u_1 < \inf_{r_2 \in R_2^c} \sup_{x_1} u_1 = \inf_{x_2} \sup_{x_1} u_1$$

REFERENCES

CASE, J.H. 1979. Economics and the competitive process. New York : New York University Press.

GREEN, J. and J.J.LAFFONT. 1979. Incentives in public decision making. Studies in public economics, vol.1. Amsterdam : North Holland Publishing Co.

LUCE, R.D. and H. RAIFFA. 1957. Games and decisions. New York : J. Wiley and Sons.

MOULIN, H. 1976."Prolongement des jeux à deux joueurs de somme nulle". Paris : Bulletin de la Société Mathématique de France 45.

MOULIN, H. 1980."On strategy-proofness and single peakedness". Public Choice 35 : 437-455.

MOULIN, H. 1981. "Prudence versus sophistication in voting strategies". Journal of Economic Theory 24, 3 : 398-412.

CHAPTER II. SOPHISTICATED BEHAVIOUR.

Prudent behaviour relies on the "complete ignorance" assumption that each player is aware of his own utility function only and ignores the other players' utility. In this chapter we explore on the contrary the "complete information" assumption that each player is aware of everybody's utility. Among non-cooperative players this allows mutual strategical anticipation of the following form: player i expects all other players j to eliminate their respective dominated strategies. Given that expectation, that everybody can perfom due to the complete information, new dominated strategies emerge for some players, and so on.

Sophisticated behaviour will be defined first in normal form games (section 1), next, equivalently, in extensive form games (section 2). Sections 3 and 4 are devoted to several applications of Kuhn's theorem.

I. SUCCESSIVE ELIMINATION OF DOMINATED STRATEGIES

Example 1. Plurality voting with ties, (Farqharson [1969]).

Among 3 candidates {a, b, c} a society {1, 2, 3} must elect one. The voting rule is plurality voting and player 1 breaks ties. In other words the strategy set (or message set) are $X_1 = X_2 = X_3 = \{a, b, c\}$ and if the agents cast the votes (x_1, x_2, x_3) the elected candidate is:

$$\pi(x_1, x_2, x_3) = x_2 \text{ if } x_2 = x_3$$
$$= x_1 \text{ if } x_2 \neq x_3$$

Suppose now that the utility of the players for the various candidates display a Condorcet effect, namely:

$$u_1(c) < u_1(b) < u_1(a)$$
$$u_2(b) < u_2(a) < u_2(c)$$
$$u_3(a) < u_3(c) < u_3(b)$$

In the 3-normal form game $(X_1, X_2, X_3, u_1o\pi, u_2o\pi, u_3o\pi)$, each player has a single lex-prudent strategy (see Problem 5 Chap. I) namely vote for his or her top candidate. Therefore lex-prudent behaviour yields the election of a. Under complete information, the situation is quite different. Observe that player 2's strategy $x_2 = b$ is dominated (by $x_2' = c$) and that strategies a and c are undominated but not equivalent:

$$u_2\pi(b, a, c) = u_2(b) < u_2(c) = u_2\pi(b, c, c)$$
$$u_2\pi(b, a, a) = u_2(a) > u_2(b) = u_2\pi(b, c, a)$$

Therefore $\mathcal{D}_2(u_2) = \{a, c\}$ and similarly $\mathcal{D}_3(u_3) = \{b, c\}$. On the other hand player 1 who arbitrates ties has a dominating strategy, namely a (we let the reader check this point). Thus if players are expected not to use any dominated strategy, the strategy sets shrink to:

$$Y_1 = \{a\} \qquad Y_2 = \{a, c\} \qquad Y_3 = \{b, c\}$$

Given these restrictions on strategies, player 2 can eliminate strategy a as being *now* dominated by c:

$u_2\pi(a, a, x_3) \leqslant u_2\pi(a, c, x_3)$ for $x_3 = b, c$, with a strict inequality at $x_3 = c$.

Similarly player 3's strategy b is *now* dominated by c:

$u_3\pi(a, x_2, b) \leqslant u_3\pi(a, x_2, c)$ for $x_2 = a, c$, with a strict inequality at $x_2 = c$.

Hence after two rounds of elimination of dominated strategies, each player is left with a single strategy, and election of c namely the worst candidate of player 1 occurs! Thus player 1's privilege of arbitrating ties is disadvantageous because it makes his strategical choice predictible at once!

Definition 1,

In the N-normal form game $G = (X_i, u_i, i \in N)$ the *successive elimination of dominated strategies* is made up of the sequences

$$X_i = X_i^o \supset X_i^1 \supset \ldots \supset X_i^t \supset X_i^{t+1} \supset \ldots \text{ all } i \in N$$

where $X_i^{t+1} = \mathcal{D}_i(u_i; X_j^t, j \in N)$.

We will say that the game G is *dominance-solvable* if there exists an integer t such that for all i, the pay-off function u_i does not depend on x_i on X_N^t :

$$\forall x_i, y_i \in X_i^t \quad \forall x_{\hat{i}} \in X_{\hat{i}}^t \quad u_i(x_i, x_{\hat{i}}) = u_i(y_i, x_{\hat{i}}) \qquad (1)$$

In this case we call X_N^t the set of *sophisticated equilibria* of G.

Clearly, if (1) holds for some t (and some $i \in N$) then $X_i^t = X_i^{t+1} = \ldots$ so that if our game is dominance-solvable the particular choice of t does not change the set X_N^t. Hence Definition 1 is consistent. To reach a sophisticated equilibrium strategy, each player i has to compute the sequences X_j^t, $t \in \mathbb{N}$, for all $j \in \mathbb{N}$ thus using fully his knowledge of the various utility functions. This computation is performed independently by each player <u>assuming that the other players do the same</u>. In this limited sense sophisticated behaviour can be said decentralized.

Dominance-solvability of G means that after finitely many elimination rounds all strategies of any player are equivalent <u>to him</u> (but not necessarily to other players: see Example 3, Chapter I). However if utility functions are all one-to-one on X_N, then the set of sophisticated equilibrium, if any, is a singleton. Thus in dominance-solvable games, sophisticated behaviour of the players is essentially deterministic.

Sophisticated equilibrium generalizes dominant strategy equilibrium in the following sense:

<u>Lemma 1</u>.

If in game G the set D of dominating strategy equilibria is non-empty, then G is dominance-solvable and D is the set of its sophisticated equilibriums.

The proof follows immediately Lemma 2, Chapter I. If only agent i is known to have a dominating strategy, then clearly $D_i(u_i)$ is the i-th component of the sophisticated equilibrium set, if the latter exists.

For normal form games no sufficient condition for dominance-solvability is known. To derive such condition it is necessary to use another representation of games, namely the extensive form.

II. EXTENSIVE FORM GAMES AND KUHN'S THEOREM

Example 2. Voting by veto.

Let $A = \{a, b, c, d\}$ be a set of four candidates among which society $N = \{1, 2, 3\}$ must pick one. The following rule is in order: starting with player 1, each player successively vetoes one among the non-vetoed candidates. The (necessarily unique) remaining candidate is elected.

Let u_1, u_2, u_3 be the players' utility functions on A. Then the situation is described by a normal form game where:

$X_1 = A$: player 1's strategy is to announce which candidate the vetoes first.

X_2 is made up of those mappings x_2 from A into A such that $x_2(\alpha) \neq \alpha$, all $\alpha \in A$. Strategy x_2 means that if player 1 vetoed α first, then player 2 vetoes next $x_2(\alpha)$.

X_3 is made up of those mappings x_3 that specify for every pair (α, β) of previously vetoed candidates, this candidate $x_3(\alpha, \beta) \in A \setminus \{\alpha, \beta\}$ that player 3 vetoes finally.

Given a triple $(x_1, x_2, x_3) \in X_1 \times X_2 \times X_3$ the corresponding elected outcome is

$$\pi(x_1, x_2, x_3) = A \setminus \{x_1, x_2(x_1), x_3(x_1, x_2(x_1))\}$$

And the normal form representation of our game is now $(X_1, X_2, X_3, u_1 o \pi, u_2 o \pi, u_3 o \pi)$. A graph theoretical representation proves to be more useful; it is the extensive form of our game:

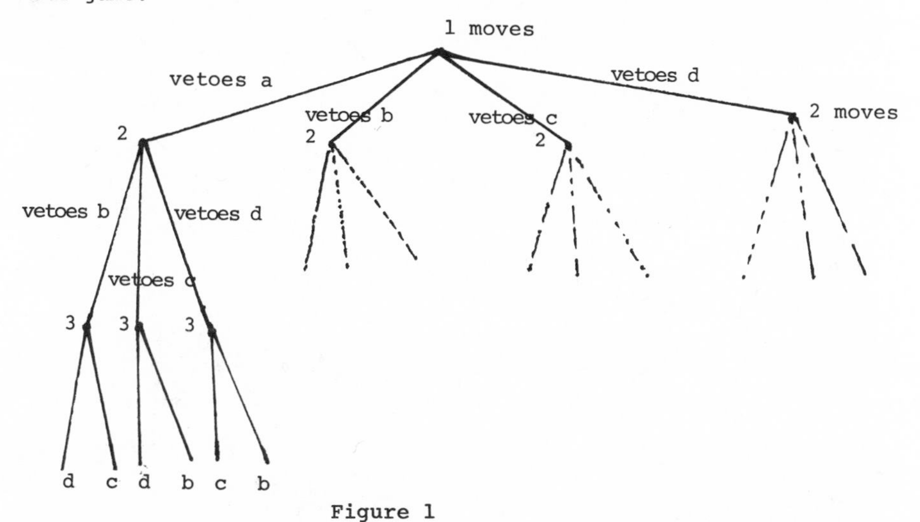

Figure 1

At each non-terminal node of this game tree is mentioned the player who freely selects a successor node. To each terminal node corresponds the finally elected candidate. We assume now that the utility functions take the particular form:

$$u_1(d) < u_1(c) < u_1(b) < u_1(a)$$
$$u_2(c) < u_2(d) < u_2(a) < u_2(b) \qquad (2)$$
$$u_3(c) < u_3(a) < u_3(b) < u_3(d)$$

Observe first that agent 3 has a dominating strategy: at each terminal node he vetoes whatever candidate he prefers less among the two candidates still alive. This defines a unique element $x_3^{\star}$ in X_3. Next remark that neither player 1 nor player 2 have a dominating strategy. In fact no strategy in X_1 is ever dominated so that $\mathcal{D}_1(u_1) = X_1$. To check this, we prove for instance that it is not a dominated strategy for player 1 to veto his top candidate a. Namely consider the following strategies $\tilde{x}_2 \epsilon X_2$, $\tilde{x}_3 \epsilon X_3$:

$$\tilde{x}_2(a) = d \qquad \tilde{x}_3(a,d) = c$$
$$\tilde{x}_2(\alpha) = a \quad \text{all } \alpha \neq a \qquad \tilde{x}_3(\alpha, a) = b \text{ if } \alpha \neq a, b$$
$$= c \text{ if } \alpha = b$$

Then $u_1\pi(a, \tilde{x}_2, \tilde{x}_3) = u_1(b) > u_1\pi(\alpha, \tilde{x}_2, \tilde{x}_3)$, all $\alpha \neq a$ so that $x_1 = a$ is not a dominated strategy of player 1.

Similarly one checks that player 2 has 2^4 undominated strategies among the 3^4 elements of X_2: a strategy x_2 belongs to $\mathcal{D}_2(u_2)$ if and only if, for all $x_1 \epsilon X_1$, $x_2(x_1)$ is not the top candidate of u_2 among $A \setminus \{x_1\}$.

The second round of elimination of dominated strategies allows player 1 to use profitably his information. Namely a completely ignorant prudent player 1 would veto his bottom candidate d ($P_1(u_1) = \{d\}$ thus enforcing election of b (since on a, b, c the utilities of player 2 and 3 generate the same ordering, so that b is elected by any pair of undominated strategies: $b = \pi(\{d\} \times \mathcal{D}_2(u_2) \times \{x_3^\star\})$).

However, player 1 can hope to do better by vetoing b. Namely after b is vetoed, players 2 and 3 face the following game:

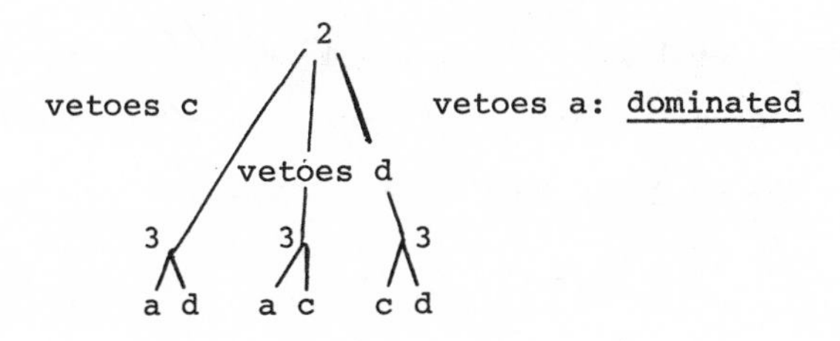

If player 2 ignores player 3's utility, he will prudently veto c ($P_2(u_2) = \{\tilde{x}_2\}$ where for all x_1, $\tilde{x}_2(x_1)$ is the bottom candidate of u_2 among $A \setminus \{x_1\}$), therefore yielding election of d, a complete failure of agent 1's ruse! If on the contrary, player 2 is informed as well, he will anticipate player 3's behaviour, therefore optimally vetoing d to enforce the election of a.

Coming back to the successive elimination of dominated strategies, we conclude that after two rounds, players 2 and 3 have deterministic strategies ($X_2^2 = \{x_2^\star\}$ is a singleton, $X_3^2 = \{x_3^\star\}$), and after three rounds player 1 as well is left with a unique strategy. The sophisticated outcome of the game is the election of a.

Definition 2.

A finite tree is a pair $\Gamma = (M,\sigma)$ where M is the finite set of nodes, where σ associates to each node its nearby predecessor, and in addition:

- There is a unique node m_o, such that $\sigma(m_o) = m_o$. We call it the origin of Γ.
- There is an integer ℓ such that $\sigma^{\ell}(m) = m_o$ for all $m \in M$. The smallest such integer is the length of Γ.

A node m such that $\sigma^{-1}(m) = \phi$ is a terminal node of Γ and their set is denoted $T(\Gamma)$. For a non-terminal node m, $\sigma^{-1}(m)$ is the set of successors of m.

Definition 3.

Let N be a finite fixed society. A N-extensive form game is defined by :

- A finite tree $\Gamma = (M,\sigma)$
- A partition $(M_i)_{i \in N}$ of $M \setminus T(M)$
- For each player $i \in N$, a utility function u_i from T(M) into R.

The partition $(M_i)_{i \in N}$ specifies which player has the move at each particular non-terminal node. If $m_o \in M_i$, then at the beginning of the play, player i must pick a successor of m_o, namely a node m_1 in $\sigma^{-1}(m_o)$. If $m_1 \in T(M)$ is a terminal node, the play is over and the corresponding pay-offs are $u_i(m_1)$, all $i \in N$. If $m_1 \notin T(M)$ is a non-terminal node, then player j such that $m_1 \in M_j$ has the move and picks a successor of m_1, namely a node $m_2 \in \sigma^{-1}(m_1)$. An so on...

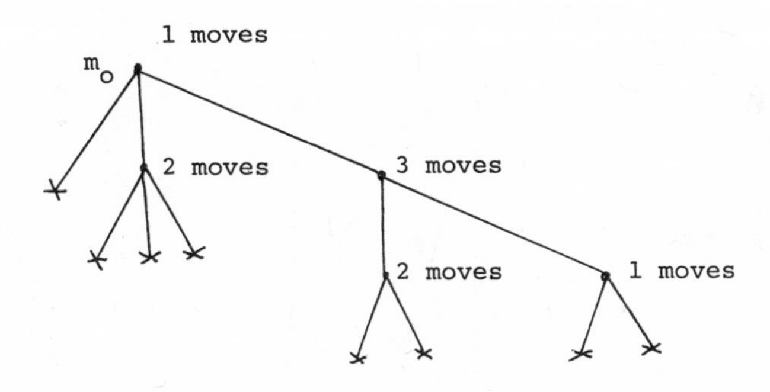

Marked nodes are terminal nodes

Figure 2: A typical 3-extensive form game

Notice that we assume that at each step of the play, the player who moves is fully informed of the current node and the entire structure of the game: this assumption is usually quoted as perfect information.

Kuhn's theorem states that finite extensive form games are in general dominance-solvable and that their sophisticated outcome is easily computed. To illustrate, consider the above voting by veto game (Example 2, Figure 1). Since in all predecessor nodes of terminal nodes, player 3 has the move, all players anticipate his behaviour from his utility function in (2). This amounts to reduce the game tree of Figure 1 as follows:

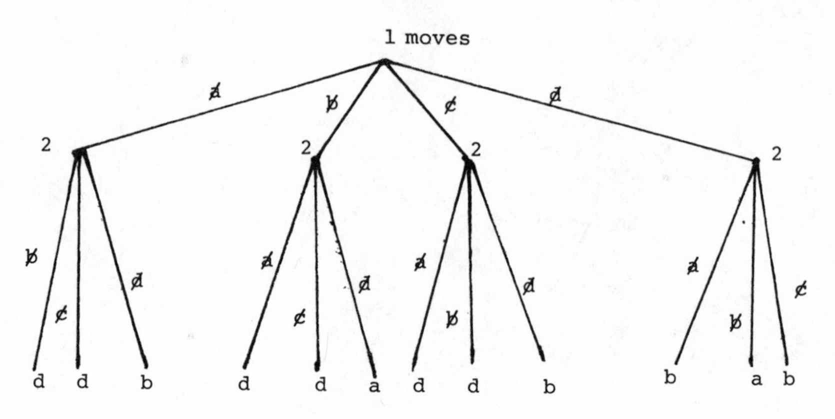

Figure 3

Since in this new tree all predecessor nodes of terminal nodes give a move to player 2, player 1 can anticipate the result of these moves from u_2 and finally face the 1-player game:

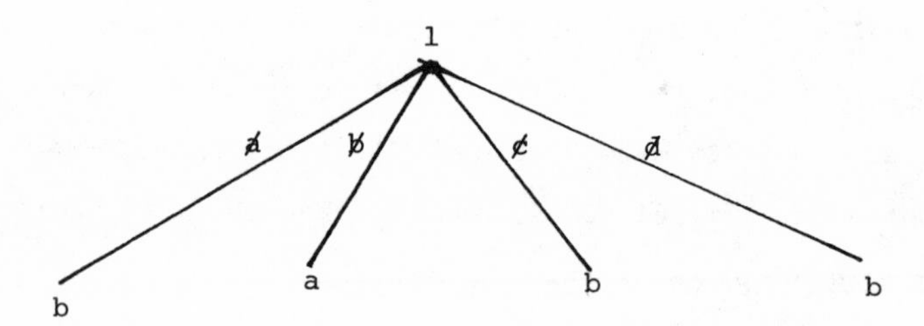

Figure 4

suggesting "veto b" as his sophisticated equilibrium strategy.

The above backward induction can be generalized to all extensive form games under a mild one-to-one assumption.

Definition 4.

Let $G = (M, \sigma; M_i, u_i; i \in N)$ be a N-extensive form game such that for any two terminal nodes $m, m' \in T(M)$ we have:

$$[\exists i \in N \quad u_i(m) = u_i(m')] \Rightarrow [\forall j \in N \quad u_j(m) = u_j(m')] \qquad (3)$$

Let $L(M)$ be the subset of those nodes m of which all successors are terminal nodes:

$$m \in L(M) \iff \phi \neq \sigma^{-1}(m) \subset T(M)$$

Then the reduced game of G is $G^\star = (M^\star, \sigma^\star, M_i^\star; u_i^\star, i \in N)$ where:

- set of nodes $M^\star = M \setminus T_o(M)$ where $T_o(M) = \{m \in T(M) / \sigma(m) \in L(M)\}$
- mapping $\sigma^\star$ = restriction of σ to $M^\star$
- terminal nodes of $(M^\star, \sigma^\star)$: $T(M^\star) = L(M) \cup \{T(M) \setminus T_o(M)\}$
- $M_i^\star = M_i \cap \{M^\star \setminus T(M^\star)\}$ hence $(M_i^\star)_{i \in N}$ partition $M^\star \setminus T(M^\star)$
- utility $u_i^\star$: $\begin{cases} \text{if } m \in T(M) \setminus T_o(M) : u_i^\star(m) = u_i(m) \\ \text{if } m \in L(M) \text{ and } m \in M_j \text{ then } u_i^\star(m) = u_i(m_j) \\ \text{where } m_j \text{ is an optimal successor to m for} \\ \text{player j: } u_j(m_j) = \sup\limits_{m' \in \sigma^{-1}(m)} u_j(m') \end{cases}$

Lemma 2.

If (M, σ) is a tree with length ℓ, then $(M^\star, \sigma^\star)$ has length $\ell - 1$.

<u>Kuhn's algorithm</u> consists of ℓ successive reductions of game G. Hence after these ℓ-reductions the game tree $(M^{\star\ell}, \sigma^{\star\ell})$ has 0-length, namely $M^{\star\ell} = \{m_o\}$ and $\sigma^{\star\ell}$ is the identity mapping. We denote by $\beta_i = u_i^{\star\ell}(m_o)$ the unique pay-off of player i in $G^{\star\ell}$.

<u>Proof of Lemma 2</u>.

Notice first that L(M) is non-empty. For if it is empty, each non-terminal node m is such that $\sigma^{-1}(m)$ contains a non-terminal node, so that we can construct an infinite sequence of distinct nodes $m_o, m_1, \ldots, m_t, \ldots$ where $m_{t+1} \in \sigma^{-1}(m_t)$, thus contradicting the fact that (M, σ) has finite length. Choose next a node m such that:

$$\sigma^{\ell}(m) = m_o, \ \sigma^{\ell-1}(m) \neq m_o \qquad (4)$$

Such a node exists by definition of ℓ. Suppose $\sigma(m) \notin L(M)$. Then there exists $m' \in M \setminus T(M)$ such that $\sigma(m') = \sigma(m)$. Picking any $m'' \in \sigma^{-1}(m')$ we get $\sigma^2(m'') = \sigma(m)$ hence $\sigma^{\ell}(m'') = \sigma^{\ell-1}(m) \neq m_o$ a contradiction. We have proved $\sigma(m) \in L(M)$. Since m is clearly a terminal node, we conclude $m \in T_o(M)$. Thus no node such that (4) holds can be in $M^{\star}$ and therefore $(M^{\star}, \sigma^{\star})$ has length at most $(\ell-1)$. It is just as easy to prove that $(M^{\star}, \sigma^{\star})$ has <u>exactly</u> length $\ell-1$.

<u>Theorem 1</u>. (Kuhn).

Let $G = (M, \sigma; M_i, u_i, i \in N)$ be an N-extensive form game and assume that (3) holds. Then, viewed as a normal form game, G is dominance-solvable and its sophisticated equilibrium pay-offs $(\beta_i)_{i \in N}$ are given by Kuhn's algorithm.

Proof.

We make precise first what we mean by the normal form of game G. Since player i has to make a strategic decision at each node of M_i a strategy x_i will be a mapping from M_i into M such that:

$$x_i(m) \in \sigma^{-1}(m) \quad \text{all } m \in M_i$$

Let X_i be the set of such mappings. A given N-uple $x = (x_i)_{i \in N}$ generates the play $m_o, m_1, \ldots, m_k$ where:

- $m_o \in M_{i_o} \Rightarrow m_1 = x_{i_o}(m_o)$
- $m_1 \in T(M) \Rightarrow k = 1$

  $m_1 \in M_{i_1} \Rightarrow m_2 = x_{i_1}(m_1)$

....

- $m_t \in T(M) \Rightarrow k = t$

  $m_t \in M_{i_t} \Rightarrow m_{t+1} = x_{i_t}(m_t)$

....

and the corresponding pay-offs are $\tilde{u}_i(x) = u_i(m_k)$. We have to prove that the game $(X_i, \tilde{u}_i, i \in N)$ is dominance-solvable and yields the same equilibrium pay-off as Kuhn's algorithm. The reduction of G into G^* (Definition 4) amounts to drop all nodes of $T_o(M)$, thus making the nodes of $L(M)$ terminal, and assigning to a node m in $L(M) \cap M_i$ the pay-off vector that would result if agent i were to take the optimal move at m. Henceforth if A_i denotes the following subset of X_i:

$$A_i = \{x_i \in X_i \ / \ \forall m \in L(M) \cap M_i = u_i(x_i(m)) = \sup_{m' \in \sigma^{-1}(m)} u_i(m')\}$$

then the N-normal form game $(A_i, \tilde{u}_i, i \in N)$ is isomorphic to the reduced game G^* (notice that if i is indifferent between two successors of $L(M)\cap M_i$, then by (3) all players are indifferent as well so that the corresponding strategies in A_i can be identified). Observe now that a strategy $x_i \in X_i \setminus A_i$ is a dominated strategy for player i (indeed x_i is dominated by a strategy $y_i \in A_i$ that coincides with x_i on $M_i \setminus L(M)$). Thus we have:

$$\mathcal{D}_i(u_i) \subset A_i \subset X_i \qquad (5)$$

We use now the following crucial Lemma:

Lemma 3. (Rochet [1980]).

Let $G = (X_i, u_i, i \in N)$ be a finite N-normal form game such that for any two outcomes $x, x' \in X_N$ we have:

$$[\exists i \in N \quad u_i(x) = u_i(x')] \Rightarrow [\forall j \in N \quad u_j(x) = u_j(x')] \qquad (6)$$

Next for all $i \in N$ let A_i be a subset of X_i such that (5) holds.

Then G is dominance-solvable if an only if $(A_i, u_i, i \in N)$ is and their sophisticated equilibrium pay-offs coincide.

Applying inductively Lemma 3 yields:

$$G \text{ d-solvable} \Longleftrightarrow G^{\star} \text{ d-solvable} \Longleftrightarrow G^{\star\star} \text{ d-solvable} \Longleftrightarrow \ldots$$

and these games all have the same sophisticated equilibrium pay-offs. Since $G^{\star \ell}$ is obviously d-solvable with equilibrium pays-offs $(\beta_i)_{i \in N}$, the proof of Theorem 1 is complete. ■

Lemma 3 is a robustness result: during the successive elimination of dominated strategies, if some players are "lazy" and do not eliminate all their dominated strategies, or if the elimination is made sequentially (player 1 eliminates first his dominated strategies, next player 2, and so on...), then the dominance-solvability is preserved and the sophisticated equilibrium pay-offs are unaffected. This result improves significantly the plausibility of sophisticated behaviour.

<u>Exercise 1</u>.

Give an example of an extensive form game which does not satisfy the one-to-one assumption (3) and of which the associated normal form game is not dominance-solvable.

<u>Problem 1</u>. <u>Proof of Lemma 3</u>. (Gretlein {1980}, Rochet {1980}).

Let $G = (X_i, u_i; i \in N)$ be a fixed finite N-normal form game such that (6) holds. If $B = \underset{i \in N}{X} B_i$ is a rectangular subset of X_N, we denote by $G(B)$ the game $(B_i, u_i; i \in N)$ and by $G(B^t) = (B_i^t, u_i; i \in N)$ the game left from $G(B)$ after t successive eliminations of dominated strategies (Definition 1). For any two rectangular subsets B, C we denote by $C \to B$ the following property:

$$\left\{\begin{array}{l} C_i \subset B_i \\ \forall x_i \in B_i \quad \exists y_i \in C_i : \forall x_{\hat{i}} \in B_{\hat{i}} \quad u_i(x_i, x_{\hat{i}}) = u_i(y_i, x_{\hat{i}}) \end{array}\right\} \text{ all } i \in N$$

1) Prove the following implications:

$$\{C \to B\} \Rightarrow \{C^1 \to B^1\}$$

2) Deduce that if $C \to B$ holds, we have the following equivalence:

G(B) is dominance-solvable <=> G(C) is dominance-solvable

If these properties hold the sophisticated equilibrium pay-offs coincide.

We proceed now to prove Lemma 3. We set r to be the cardinality of X_N and assume that Lemma 3 holds for any game whose outcome set has cardinality strictly less than r.

We fix a rectangular subset A of X_N such that (5) holds. If $|A| = r$ there is nothing to prove so that we assume $|A| \leqslant r-1$. We set $B = A^1 \cup X^2$ (where $X^2 = \underset{i \in N}{X} X_i^2$ and $A^1 = \underset{i \in N}{X} A_i^1$) and observe that:

$$A^1 \subset B \subset A$$

Thus by the induction assumption we have: G(B) is dominance-solvable if and only if G(A) is and their sophisticated equilibrium pay-offs coincide.

3) Setting $C = (A^1 \cap X^1) \cup X^2$ prove that $C \rightarrow B$

4) Remark that $X^2 \subset C \subset X^1$ and conclude .

III. <u>TWO-PERSON ZERO-SUM GAMES</u>

We first need an auxiliary result.

<u>Lemma 4</u>.

Let $G = (X_1, X_2; u_1)$ be a two person zero-sum game with finite strategy sets. Then a sophisticated equilibrium of G is a saddle point of u_1. Hence a dominance-solvable two person zero-sum game has a value.

Proof.

Given two subjects $Y_1 \subset X_1$, $Y_2 \subset X_2$ we denote by $S(Y_1, Y_2)$ the-possibly empty-set of saddle-points of the game (Y_1, Y_2, u_1) Next we set $Z_i = \mathcal{D}_i(u_i; Y_1, Y_2)$. Let x_1, x_2 be a saddle point of game (Z_1, Z_2, u_1) and suppose that (x_1, x_2) is not a saddle-point of game (Y_1, Y_2, u_1) : for instance there exists $y_1 \in Y_1$ such that:

$$u_1(x_1, x_2) < u_1(y_1, x_2) \qquad (7)$$

We set:

$$Y_1(y_1) = \{y_1' \in Y_1 \,/\, \forall y_2 \in Y_2 \quad u_1(y_1, y_2) \leq u_1(y_1', y_2)\}$$

We claim that $Y_1(y_1)$ has a non-empty intersection with Z_1. For instance if z_1 reaches the maximum of ϕ_1 over $Y_1(y_1)$ where:

$$\phi_1(y_1') = \sum_{y_2 \in Y_2} u_1(y_1', y_2)$$

then one checks that z_1 cannot be dominated in (Y_1, Y_2, u_1). Therefore z_1 is in $Y_1(y_1) \cap Z_1$ and from (7) we derive:

$$u_1(x_1, x_2) < u_1(z_1, x_2)$$

a contradiction of our assumption that (x_1, x_2) is a saddle-point of (Z_1, Z_2, u_1).

We have just proved:

$$S(Z_1, Z_2) \subset S(Y_1, Y_2)$$

Applying the above inclusion iteratively yields :

$$S(X_1, X_2) \supset S(X_1^1, X_2^1) \supset \ldots \supset S(X_1^t, X_2^t)$$

Now if G is dominance-solvable, there is an integer t such that all strategies in X_i^t are equivalent to player i given that player j' strategies are restricted to X_j^t (Definition 1). For such a t we have in particular:

$$S(X_1^t, X_2^t) = X_1^t \times X_2^t = \{\text{sophisticated equilibriums of } G\}$$

∎

This concludes the proof of Lemma 4.

Corollary of Theorem 1 and Lemma 4.

Every finite extensive two-person zero-sum game $(M, \sigma; M_1, M_2, u_1)$ has a value. Each player has at least one optimal strategy.

Namely property (3) follows from $u_2 = - u_1$. An illustration of this Corollary is Zermelo's theorem: in chess either White can force a win, either Black can force a win or both players can force a draw. Chess is indeed a finite extensive two-person zero-sum game (with three pay-offs White wins, Draw, Black wins) by the rule that declares draw when the same position on the board appears three times.

Example 3. A game of Nim.

For any integer $n \geq 1$ denote by G_n^1 the following two-person zero-sum game. If $n = 1$ then G_1^1 is the trivial game where player 1 wins.

If $n \geq 2$ player 1 must pick a number n_1; $1 \leq n_1 \leq n-1$. Next player 2 chooses either one of n_1 or $n_2 = n - n_1$. Call n_i the choice of player 2: then the next step is to play $G_{n_i}^2$, the same game starting at n_i where player 2 plays first. In particular if either one of n_1, n_2 is 1, player 2 wins by picking it.

This defines an extensive form game of length $(n-1)$. By the above Corollary it has a value therefore the set $\mathbb{N}$ of integers can be partitionned as $\mathbb{N} = N_1 \cup N_2$ where N_i is the set of those n such that player i can force a win in G_n^1. By an obvious induction argument on n we get that our partition satisfies $1 \in N_1$ and moreover:

$$\begin{cases} \forall n \in N_1 \ (n \geq 2) \Rightarrow \exists n_1,\ 1 \leq n_1 \leq n-1;\ n_1 \in N_2 \text{ and } n - n_1 \in N_2 \\ \forall n \in N_2 \quad \forall n_1,\ 1 \leq n_1 \leq n-1:\ n_1 \in N_1 \text{ and/or } n - n_1 \in N_1 \end{cases} \tag{8}$$

We let the reader check that these conditions together give:

$$N_1 = \{n / \exists p \geq 1 \quad n = 5p \quad \text{or } n = 5p - 4 \quad \text{or } n = 5p - 1\}$$

Consider now the games Γ_n^1, Γ_n^2 with the same rules except that the player picking a 1 looses, in other words in Γ_1^i player j, $j \neq i$, wins. The same argument shows the existence of a partition $N = M_1 \cup M_2$ such that $1 \in M_2$ and (8) holds with M_i instead of N_i, $i = 1, 2$. This in turn characterizes M_1 as the set of <u>even</u> integers.

As a matter of exercize the reader will describe player 1's winning strategies in G_n^1, $n \in N_1$ and in Γ_n^1, n even and player 2's winning strategies in G_n^1, $n \in N_2$ and in Γ_n^1, n odd.

<u>Problem 2</u>. <u>The Marienbad game</u>.

We fix an integer p, $p \geq 1$ and for all p-uples $\underline{n} = (n_1, \ldots, n_p)$ where $n_1, \ldots, n_p$ are integers, possibly zero, we define: $G_{\underline{n}}^1$, $G_{\underline{n}}^2$ by the following induction argument : Given $\underline{n} \neq (0, 0 \ldots, 0)$ call $\underline{n}'$ a <u>successor</u> of $\underline{n}$ if there exists k, $1 \leq k \leq p$ such that:

$$\begin{cases} n'_{k'} = n_{k'} \text{ for all } k',\ 1 \leq k' \leq p \text{ and } k' \neq k \\ 0 \leq n'_k < n_k \end{cases}$$

If $\underline{n} = (0, .., 0)$ then in $G^i_{\underline{n}}$ player i wins. If $\underline{n} \neq (0, .., 0)$ then $G^i_{\underline{n}}$ is played as follows: first player i picks a successor $\underline{n}'$ of $\underline{n}$. If $\underline{n}' = (0, .., 0)$ then player j, $j \neq i$, wins the play, otherwise the game $G^j_{\underline{n}'}$ starts: player j picks a successor $\underline{n}''$ of $\underline{n}'$, if $\underline{n}'' = (0, .., 0)$ player i wins, otherwise $G^i_{\underline{n}''}$ starts, and so on...

Physically the game amounts to set p rows of respectively $n_1, .., n_p$ matches and ask each player successively to strictly reduce (by an arbitrary amount) exactly one row still alive. The player who kills the last row alive looses.

1) Prove the existence of a partition:

$$[\mathbb{N} \cup \{0\}]^p = N_1 \cup N_2$$

such that if $\underline{n}$ belongs to N_i, player i can force a win in $G^1_{\underline{n}}$. Characterize this partition by conditions similar to (8).

2) For all $\underline{n} = (n_1, .., n_p)$ denote by

$$n_k = \alpha^k_\ell \alpha^k_{\ell-1} \cdots \alpha^k_1 \alpha^k_o \quad , 1 \leq k \leq p$$

the diadic development of n_k, where ℓ is an upper bound of the number of digits needed for $n_1, \ldots, n_p$ (thus some among α^1_ℓ , $\ldots, \alpha^p_\ell$ might be zero, not all of them). Next denote

$$a_o = \sum_{k=1}^{p} \alpha^k_o, \ \ldots, \ a_\ell = \sum_{k=1}^{p} \alpha^k_\ell$$

Then prove that $N_2 = M_2 \cup P_2$ where

$$M_2 = \{\underline{n} / \forall j, \ 0 \leq j \leq \ell, \ a_j \text{ is even and } \exists j \ , \ 1 \leq j \leq \ell, \ a_j > 0\}$$

$$P_2 = \{\underline{n} / \forall j, \ 1 \leq j, \ a_j = 0 \text{ and } a_o \text{ is odd}\}$$

Problem 3. A topological duel, (Choquet).

Let E be a metric space. We denote by O the set of non-empty open subsets of E. Our game works as follows:

In Step 1 player 1 picks an $A_1 \in O$

In Step 2 player 2 picks an $A_2 \in O$ with the only constraint $A_2 \subset A_1$.

...

In step t a player (1 for odd t, 2 for even t) picks an $A_t \in O$ with the only constraint $A_t \subset A_{t-1}$

..., and so on undefinitely.

We say that player 1 wins the play if

$$\bigcap_{t=1}^{\infty} A_t \neq \emptyset$$

If this intersection is empty then we say that player 2 wins the play.

1) Prove that if E is a complete metric space, player 1 can force a win.

2) Prove that if $E = \mathbb{Q}$ player 2 can force a win.

3) Do you think that for any E our game has a value?

IV. OTHER APPLICATIONS TO KUHN'S THEOREM

So far we have considered situations where all players are at the same information level (complete ignorance - Chapter I -, or complete information - this Chapter). In several economic applications (for instance in oligopolistic competition with a "dominant" firm) a asymmetrical allocation of information arises naturally. With the help of Kuhn's theorem we explore the simplest model of this sort: <u>the leader-follower behaviour in two person games</u>.

Given a two person game (X_1, X_2, u_1, u_2) we will denote by BR_i the graph of player i' best reply correspondence:

$$(x_1, x_2) \in BR_1 \iff u_1(x_1, x_2) = \sup_{y_1 \in X_1} u_1(y_1, x_2)$$

(with a symmetrical definition for BR_2)

<u>Definition 5.</u>

We say that (x_1, x_2) is a i-<u>stackelberg equilibrium</u> of game (X_1, X_2, u_1, u_2) if:

$$(x_1, x_2) \in BR_j \text{ and } u_i(x_1, x_2) = \sup_{(y_1, y_2) \in BR_j} u_i(y_1, y_2) \quad (9)$$

where $i, j = 1, 2$ and $i \neq j$

We interpret a 1-stackelberg equilibrium within the following scenario: player 1 - the leader - is informed of both utility functions u_1, u_2 and uses this information to anticipate player 2's reactions; player 2 - the follower - takes player 1's strategy as exogeneously given (he typically ignores player 1's utility) and reacts by maximizing his pay-off, given that player 1's strategy is fixed. Thus player 1 has the first move and anticipates that player 2 will use a best reply to x_1: therefore he optimally solves problem (9).

Remark that if player 2 has several best reply strategies to x_1, it is assumed by (9) that he will break this tie in favour of u_1. This simplifying assumption has little impact on the subsequent argument.

Lemma 5.

Let $G = (X_1, X_2, u_1, u_2)$ be a finite two person game where u_1, u_2 are both one-to-one on $X_1 \times X_2$. Then there is a unique 1-stackelberg equilibrium, denoted $(\bar{x}_1, \bar{x}_2)$. Consider next the game $\tilde{G} = (X_1, X_2^{X_1}, \tilde{u}_1, \tilde{v}_1)$ given by:

$$\begin{cases} X_2^{X_1}\text{, with current element } \eta \text{ is the set of mappings} \\ \text{from } X_1 \text{ into } X_2 \\ \forall x_1 \in X_1 \quad \forall \eta \in X_2^{X_1}: \tilde{u}_i(x_1, \eta) = u_i(x_1, \eta(x_1)) \end{cases}$$

Then the game G is dominance-solvable with a unique sophisticated equilibrium $(\bar{x}_1, \bar{\eta})$ where $\bar{\eta}$ is the best reply strategy of player 2, and $\bar{\eta}(\bar{x}_1) = \bar{x}_2$.

Proof.

Existence and uniqueness of the 1-stackelberg equilibrium follows from the one-to-one character of u_1 on $X_1 \times X_2$. The game $\tilde{G}$ is the normal form of the extensive form game where player 1 chooses first a strategy in X_1, next player 2, knowing of player 1's choice selects his strategy in X_2. In $\tilde{G}$ the best reply strategy $\bar{\eta}$ is a dominating strategy of player 2. Namely for all $x_1 \in X_1$ and all $\eta \in X_2^{X_1}$ we have:

$$\tilde{u}_2(x_1, \bar{\eta}) = u_2(x_1, \bar{\eta}(x_1)) = \sup_{x_2 \in X_2} u_2(x_1, x_2)$$
$$\geq u_2(x_1, \eta(x_1)) = \tilde{u}_2(x_1, \eta)$$

Our one-to-one assumption defines $\bar{\eta}$ unambiguously. Before the second round of elimination of dominated strategies, player 1 faces the game $(X_1^1, \{\bar{\eta}\}, u_1, u_2)$ where his dominating strategy is the - unique - strategy $x_1^\star$ such that:

$$\tilde{u}_1(x_1^\star, \bar{\eta}) = u_1(x_1^\star, \bar{\eta}(x_1^\star)) \geq u_1(x_1, \bar{\eta}(x_1)) = \tilde{u}_1(x_1, \bar{\eta})$$
$$\text{all } x_1$$

By the one-to-one character of u_2, the graph of $\bar{\eta}$ equals BR_2. Therefore $(x_1^\star, \bar{\eta}(x_1^\star))$ is the 1-stackelberg equilibrium of our game, so that $x_1^\star = \bar{x}_1$ and $\bar{\eta}(x_1^\star) = \bar{x}_2$.

∎

Remark that under the usual topological assumptions (X_1, X_2 compact, u_1, u_2 continuous) existence of a i-stackelberg equilibrium, i = 1, 2, is guaranteed. However Lemma 5 cannot be generalized straightforwardly.

Example 4. A Rawlsian voting procedure.

Let A = {1, 2, ..., 7} be a set of 7 candidates among which 2 players must select one. Each player must rank these 7 candidates to complete his message. Hence a strategy x_i is a one-to-one mapping from A onto {1, 2, ..., 7} where it is understood that $x_i(a) = 1$ means that candidate a is said by player i to be his top candidate (of course all lies are permitted). We denote by Z the common strategy set.

Given a pair of strategies $(x_1, x_2) \in Z \times Z$ the elected candidate is $\pi(x_1, x_2)$ where:

$$\begin{cases} \pi(x_1, x_2) = \inf \{a / a \in R(x_1, x_2)\} \\ R(x_1, x_2) = \{a \in A / \max[x_1(a), x_2(a)] = \min\limits_{b \in A} \max[x_1(a), x_2(a)]\} \end{cases}$$

Clearly $R(x_1, x_2)$ contains no more than two candidates. This voting rule is intended to elicit a candidate whose worst score among the two voters is as high as possible. Given that the voter-players can cast any ranking, a strategic game emerges. Let us denote by $u_1, u_2 \in Z$ the <u>true</u> opinion of player 1, 2. Then the voters face the normal form game:

$$(Z, Z, -u_1 o\pi, -u_2 o\pi) \qquad (10)$$

(ranking are the opposites of utility functions).

We compute now the 1-stackelberg equilibrium of game (10). Let us fix a strategy $x_1 \in Z$ of player 1. Observe first that for all $x_2 \in Z$:

$$\pi(x_1, x_2) = a \Rightarrow x_1(a) \leqslant 4 \qquad (11)$$

In words by sending message x_1, player 1 vetoes candidates $x_1^{-1}(7)$, $x_1^{-1}(6)$, $x_1^{-1}(5)$. This is true because at least one candidate $\tilde{a}$ in $x_1^{-1}(\{4, 3, 2, 1\})$ must be such that $x_2(\tilde{a}) \leqslant 4$ and therefore

$$\max(x_1(\tilde{a}), x_2(\tilde{a})) \leqslant 4 < \max(x_1(b), x_2(b)) \text{ for all } b$$

in $x_1^{-1}(\{7, 6, 5\})$

Next observe that, given x_1, player 2 can force the election of any candidate in $x_1^{-1}(\{4, 3, 2, 1\})$

$$\forall a \in A \quad [x_1(a) \leqslant 4] \Rightarrow [\exists x_2 \in Z \quad \pi(x_1, x_2) = a] \qquad (12)$$

For instance, to force the election of $x_1^{-1}(4)$, player 2 sends the following message:

	x_1	x_2
top rank:	a	d^*
	b	e
	c	f
	d^*	g
	e	a
	f	b
bottom rank:	g	c

From (11) and (12) we deduce that any best reply strategy x_2 to x_1 is such that:

$$\pi(x_2, x_1) = a \text{ where } u_2(a) = \inf\{u_2(b)/x_1(b) \leq 4\} \qquad (13)$$

This implies in particular:

$$\forall (x_1, x_2) \in BR_2 \qquad u_2(\pi(x_1, x_2)) \leq 4$$

(since $\{b/x_1(b) \leq 4\}$ has cardinality 4).

Therefore, by acting as a leader, player 1 will necessarily force the election of a candidate within $A_2 = u_2^{-1}(\{1,2,3,4\})$. Actually he can force the election of any candidate in A_2 by sending an appropriate message. For instance, to force the election of $a = u_2^{-1}(3)$, player 1 just announces the following message:

	u_2	x_1
top:	b	a
	c	e
	a	f
	d	g
	e	b
	f	c
bottom:	g	d

which raises a best reply x_2 such that (by (13)) $\pi(x_1, x_2) = a$.

Finally we conclude that Player 1's pay-off (expressed in ranks) at any 1-stackelberg equilibrium is:

$$S_1 = \inf\{u_1(a)/u_2(a) \leqslant 4\}$$

As the role of both players is symmetrical in our voting rule, we get similarly player 2's pay-off (rank) at any 2-stackelberg equilibrium:

$$S_2 = \inf\{u_2(a)/u_1(a) \leqslant 4\}$$

An interesting corollary is that in general <u>the ranks S_1 and S_2 are not compatible</u>, that is to say:

there is no a such that $u_1(a) \leqslant S_1$ <u>and</u> $u_2(a) \leqslant S_2$ (14)

More precisely, a pair (u_1, u_2) <u>either</u> satisfies (14) <u>or</u> is such that:

$$\begin{cases} \text{exactly one candidate } a^\star \text{ is such that } u_1(a^\star) \leqslant 4 \\ \text{and } u_2(a^\star) \leqslant 4. \text{ Then } S_i = u_i(a^\star) \text{ for } i = 1, 2 \end{cases}$$

When (u_1, u_2) is such that (14) holds, the game (10) displays a typical <u>struggle for the leadership</u>: as long as players are mutually aware of their preferences, it always pays to have the first move, that is to force the other player into the follower's position. We shall study in more detail this specific conflict in Chapter VI Section 4, in connection with the geometry of threats. See also Example 2 and Lemma 2, Chapter III.

As a conclusion to Chapter II we illustrate on several examples and problems some technical or conceptual difficulties surrounding the notion of sophisticated behaviour.

When the strategy sets X_i, $i \in N$ are infinite - say compact sets - the very notion of sophisticated behaviour is quite difficult to formalize. One reason is that along the successive elimination of dominated strategies (Definition 1), compactness of the sets X_i^t is not guaranteed (see Problem 2 Chapter I). Moreover convergence of the sequences X_i^t, $t \in \mathbb{N}$ to a subset of equivalent strategies may hold only in the limiting sense, thus raising more topological difficulties... Last but not least the one-to-one assumption (3), which is crucial to the proof of Kuhn's theorem, becomes difficult to sustain. These difficulties are illustrated by our next example as well as Problem 4, 5 below.

<u>Example 5</u>. <u>Dividing an inflated dollar</u>. (Dutta-Gevers [1981]).

Agents {1, 2, ..., n} divide a dollar among themselves, according to the following rule:

<u>Step 1</u>: player 1 proposes a sharing $x^1 = (x_1^1, \ldots, x_n^1)$ where

$\sum_{i=1}^{n} x_i^1 = 1$ and $x_i^1 \geqslant 0$, all $i \in N$. Then agents 2, ..., n each have the option to accept x^1 or reject it. If all agents agree on x^1, it is done. If at least one agent rejects x^2, then we go to

Step 2: player 2 submitts a proposal x^2 to the unanimous approval of the other agents. If rejected we go to Step 3 where agent 3 makes a proposal, and so on... If step n is ever reached, and player n's proposal is rejected, then the whole procedure starts again, with a proposal by player 1, and so on...

We assume that the initial dollar is depreciated at each period by a discount factor τ, $0 < \tau < 1$. Thus at period 2 it remains $\delta = 1 - \tau$ dollar to be shared, δ^2 at period 3 and so on... Of course in the event that the division procedure goes on undefinitely, each player ultimately makes a zero profit.

Assuming existence of a sophisticated equilibrium outcome $x^* = (x_1^*, \ldots, x_n^*)$ (for a proper definition and an existence proof, see Rubinstein [1980]) we make the usual anticipation argument to actually compute x^*. Suppose player 1 at step 1 proposes x^1. He knows that player i, $i \neq 1$ anticipates from Step 2, if any, an equilibrium outcome $(\delta x_n^*, \delta x_1^*, \delta x_2^*, \ldots, \delta x_{n-1}^*)$. Namely the game starting at Step 2 is obtained from our original game by accurately permuting the players. Thus in order for proposal x^1 to be agreed upon by agents 2, ..., n, it is necessary that:

$$x_2^1 \geqslant \delta x_1^*, \; x_3^1 \geqslant \delta x_2^*, \; \ldots, \; x_n^1 \geqslant \delta x_{n-1}^* \qquad (15)$$

By assumption, if proposed $x^{\star}$ will be agreed upon therefore:

$$x_2^{\star} \geqslant \delta x_1^{\star}, \ldots, x_n^{\star} \geqslant \delta x_{n-1}^{\star} \qquad (16)$$

Pick now an x_1^1 such that:

$$x_1^1 < 1 - [\delta x_1^{\star} + \delta x_2^{\star} + \ldots + \delta x_{n-1}^{\star}]$$

then there exists a feasible sharing $(x_1^1, x_2^1, \ldots, x_n^1)$ making all inequalities in (15) strict and therefore agreed upon by agents 2, ..., n, if proposed. Hence inequality $x_1^{\star} < 1 - [\delta x_1^{\star} + \ldots + \delta x_{n-1}^{\star}]$ would violate optimality of $x^{\star}$ (this is the Nash equilibrium property of the sophisticated equilibriums - Theorem 1 Chapter III below -). Thus we have:

$$1 - [\delta x_1^{\star} + \ldots + \delta x_{n-1}^{\star}] = (1-\delta) + \delta x_n^{\star} \leqslant x_1^{\star} \qquad (17)$$

Combining (16) and (17) yields for all i = 1, ..., n

$$x_i^{\star} \geqslant \delta^{i-1} \cdot x_1^{\star} \geqslant (1-\delta)\delta^{i-1} + \delta^i x_n^{\star} \geqslant (1-\delta)\delta^{i-1} + \delta^n x_i^{\star}$$

which implies

$$x_i^{\star} \geqslant \frac{\delta^{i-1}}{1+\delta+\ldots+\delta^{n-1}} \quad \text{all } i = 1, \ldots, n$$

As the right-hand terms in the above inequalities sum up to 1, we conclude that all are equalities. Notice that when δ goes to 1, the limit outcome is the egalitarian sharing (each agent gets $\frac{1}{n}$); however for $\delta = 1$ the whole strategical argument collapses.

Our next example uses a genuine solution concept to describe properly the non-cooperative behaviour of completely informed agents. Namely after elimination of dominated strategies the remaining restricted game is inessential.

Example 6. The Steinhaus method to share a cake.

Let $[0, 1]$ be a non-homogeneous cake to be divided among two players. Utility of player 1 for a share $A \subset [0, 1]$ is worth to him:

$$v_1(A) = \int_A (\frac{3}{2} - x)dx$$

whereas utility of player 2 for a share $B \subset [0, 1]$ is worth to him:

$$v_2(B) = \int_B (\frac{1}{2} + x)dx$$

When time runs from $t = 0$ to $t = 1$ a knife is moved at speed 1 from $x = 0$ to $x = 1$. At any time both players can stop it; if the knife is stopped at time t by player i, this player gets the share $[0, t]$ whereas the other player gets $[t, 1]$. Thus the strategy sets are $X_1 = X_2 = [0, 1]$ where strategy x_i means that player i will stop at time $t = x_i$ unless the other player did so before that time. The utility functions are given by:

$$u_1(x_1, x_2) = \begin{cases} v_1([0, x_1]) = \frac{3}{2}x_1 - \frac{x_1^2}{2} & \text{if } x_1 \leq x_2 \\ v_1([x_2, 1]) = 1 - \frac{3}{2}x_2 + \frac{x_2^2}{2} & \text{if } x_2 < x_1 \end{cases}$$

$$u_2(x_1, x_2) = \begin{cases} v_2([x_1, 1]) = 1 - \frac{1}{2}x_1 - \frac{1}{2}x_1^2 & \text{if } x_1 \leq x_2 \\ v_2([0, x_2]) = \frac{1}{2}x_2 + \frac{1}{2}x_2^2 & \text{if } x_2 < x_1 \end{cases}$$

The strategy t_i, $i = 1, 2$ defined by:

$$v_i([0, x_i]) = v_i([x_i, 1])$$

is the unique prudent strategy of player i. Namely we have:

$$\varphi_i(x_i) = \inf_{x_j \in [0,1]} u_i(x_1, x_2) = \inf \{v_i([0, x_i]), v_i([x_i, 1])\}$$

so that φ_i reaches its maximum when both shares $[0, t_i]$ and $[t_i, 1]$ are indifferent to player i. By our specific choice of v_i we have clearly

$$t_1 < t_2$$

Thus prudent behaviour implies that player 1 gets the share $[0, t_1]$ hence his secure utility level $\alpha_1 = v_1([0, t_1])$, whereas player 2 is more lucky: $v_2([t_1, 1]) > v_2([t_2, 1]) = \alpha_2$. In fact the prudent outcome allocates the whole surplus available from (α_1, α_2) to player 2, namely:

$$v_2([t_1, 1]) = \max\{u_2(x_1, x_2)/u_1(x_1, x_2) \geqslant \alpha_1\}$$

We shall say in Chapter V that this outcome is the best imputation for player 2.

The first round of elimination of dominated strategies give:

$$\mathcal{D}_i(u_i) = [t_i, 1]$$

namely a strategy x_i that would stop the knife before t_i is dominated by the prudent strategy t_i (as is easily checked). Suppose now that player 1 is informed of player 2's utility for the cake and therefore can compute t_2. Given that player 2 will not stop the knife before t_2, a second round of elimination of dominated strategies allows him to drop the strategies in $[t_1, t_2[$. Indeed for all $x_1 \in [t_1, t_2[$ we have:

$$v_1([0, x_1]) = u_1(x_1, x_2) < u_1(t_2, x_2) = v_1([0, t_2]), \text{ all } x_2 \in [t_2, 1]$$

Thus $X_1^2 = [t_2, 1]$ and no further elimination of dominated strategies is possible for either player:

$$X_1^t = X_2^t = [t_2, 1] \quad \text{all } t \geqslant 2$$

Since no two strategies in X_i^t are equivalent to player i in $X_1^t \times X_2^t$ we conclude that our game is <u>not</u> dominance-solvable. However the restricted game $([t_2, 1]), [t_2, 1], u_1, u_2)$ is inessential. Namely both players have there the same prudent strategy $x_1 = x_2 = t_2$:

$$\tilde{\alpha}_1 = \inf_{x_2 \in [t_2, 1]} u_1(t_2, x_2) = v_1([0, t_2]) > \inf_{x_2 \in [t_2, 1]} u_1(x_1, x_2)$$

$$= v_1([x_1, 1]), \text{ all } x_1 > t_2$$

$$\tilde{\alpha}_2 = \inf_{x_1 \in [t_2, 1]} u_2(x_1, t_2) = v_2([t_2, 1]) > \inf_{x_1 \in [t_2, 1]} u_2(x_1, x_2)$$

$$= v_2([x_2, 1]), \text{ all } x_2 > t_2$$

Furthermore $(\tilde{\alpha}_1, \tilde{\alpha}_2)$ is a Pareto optimal utility vector (Definition 5 Chapter I) as it is clear on Figure 5. We conclude from Theorem 1 Chapter I that $x_1 = x_2 = t_2$ is an optimal strategy for both players in the 2-reduced game, which makes (t_2, t_2) the predicted outcome when player 1 has complete information on player 2's utility. Notice that $(\tilde{\alpha}_1, \tilde{\alpha}_2) = (v_1([0, t_2]), v_2([t_2, 1])$ is Pareto optimal as well in the initial game and allocates the whole surplus to player 1:

$$v_1([0, t_2]) = \max\{u_1(x_1, x_2)/u_2(x_1, x_2) \geq \alpha_2\}$$

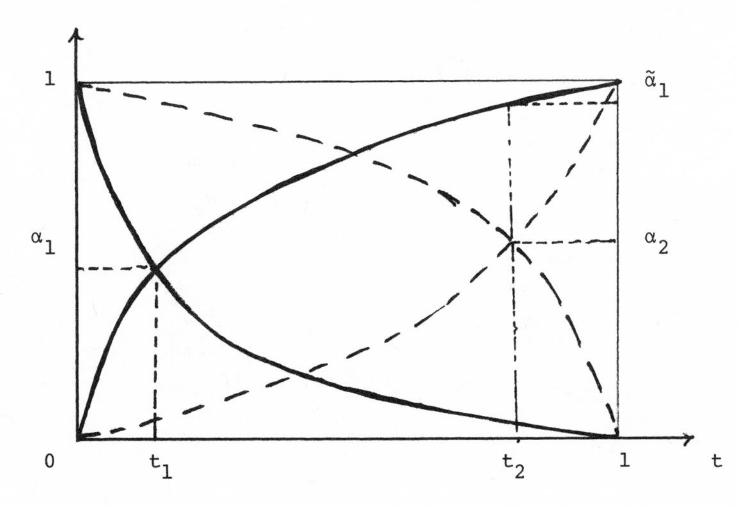

Figure 5

The two curves crossing at (t_1, α_1) are the graphs of $v_1([0, t])$ and $v_1([t, 1])$. The two dotted curves crossing at (t_2, α_2) are the graphs of $v_2([0, t])$ and $v_2([t, 1])$.

Exercise 2.

If the knife is now moved from $x = 1$ to $x = 0$, which non-cooperative decentralized behaviour emerges?

Exercise 3. Auction of an indivisible good.

Consider the sealed bid first price auction of an indivisible good (Example 2 Chapter I). Perform the successive elimination of dominated strategies (two rounds are enough). Prove that the game is not dominance-solvable and the reduced game (after the successive elimination) is not inessential. Remark however that player 1 has a unique prudent strategy in the reduced game thus raising a "natural" non-cooperative outcome of the game.

Problem 4. The Divide-and-Choose method,

Now player 1 picks a number $x_1 \epsilon [0, 1]$, next player 2 chooses either one of $[0, x_1]$ or $[x_1, 1]$ as his share (and player 1 gets the remaining share). The same utility functions as in Example 5 describe the players' preferences.

1) State the normal form of the game.

2) Prove that each player has a unique prudent strategy and that prudent behaviour allows player 2 to keep all the surplus.

3) Assuming that player 2, when indifferent among two shares, chooses so as to favour player 1, prove that the game is dominance-solvable and that its sophisticated outcome gives all the surplus to player 1.

Can you suggest a modification of Definition 1 that allows to relax the "ad hoc" assumption and predicts essentially the same behaviour?

4) What are the i-stackelberg equilibria of this game? Interpret the results.

Problem 5. Two person bargaining with a discount rate. (Binmore, 1980).

Two agents bargain on a pay-off subset $A \subset \mathbb{R}^2_+$ which takes the following form

$$A = \{(z_1, z_2) \ / \ 0 \leqslant z_2 \leqslant \theta(z_1), \ 0 \leqslant z_1\}$$

where θ is a differentiable function on $[0, 1]$ such that:

$$\left.\begin{array}{ll} \theta(1) = 0 & \theta'(t) < 0 \\ \theta(0) = 0 & \theta''(t) < 0 \end{array}\right\} \quad \text{all } t, \ 0 \leqslant t \leqslant 1$$

The method is the same as in Example 5: at Step 1 player 1 proposes an outcome $x^1 \epsilon A$. If player 2 rejects x^1, he proposes

$x^2 \in \delta A$, where δ, $0 < \delta < 1$ is the discount factor. If player 1 rejects x^2 he suggests $x^3 \in \delta^2 A$, and so on... If agreement is reached at some step t, the coordinates of x^t are the final pay-offs of both players. Otherwise both players get a zero pay-off.

1) Prove that at the sophisticated equilibrium outcome, agreement is reached at step 1 on a Pareto optimal vector $x(\delta)$ of A. Characterize x^* with the help of function θ.

2) Prove that as δ goes to 1, $x_1(\delta)$ decreases (and $x_2(\delta)$ increases). Prove that the limiting outcome $x(1)$ is the Nash bargaining solution of A, i.e. the solution of the following problem:

$$\max_{z \in A} \; z_1 . z_2$$

<u>Problem 6</u>. <u>Public decisions mechanisms with side-payments</u>. (Moulin [1981]).

A society $N = \{1, 2, \ldots, n\}$ must pick one public decision among a finite set $A = \{a, b, c, \ldots\}$. One private good (money) allows side-payments and each player's utility is quasi-linear (as in Problem 4 Chapter I): player i's utility is described by a vector $u_i \in \mathbb{R}^A$. If the public decision a is taken and i receives a monetary transfer t_i (which might be positive or negative) his final utility level is:

$$u_i(a) + t_i$$

A <u>decision</u> $(a; t)$ is made of a public decision $a \in A$ and a vector $t = (t_1, \ldots, t_n)$ of monetary transfers such that $t_1 + \ldots + t_n = 0$. We denote by $\mathcal{D} \subset A \times \mathbb{R}^n$ the set of such decisions.

1) We will consider two distinct mechanisms. In the first one the agents successively propose a decision to unanimous approval by the others. An agent whose proposal has been rejected has no longer veto power and no monetary transfer can be levied on him anymore.

More precisely:

Agent 1 proposes first a decision $d^1 = (a^1, t^1) \in \mathcal{D}$ to the unanimous approval of the other agents. If every other agent accepts d^1 then it is the final decision. If at least one agent rejects d^1 then agent 2 has to propose a decision $d^2 = (a^2, t^2) \in \mathcal{D}$ with the only constraint $t_1^2 = 0$.

Decision d^2 is submitted to the unanimous approval of agents $\{3, 4, \ldots, n\}$. If it is rejected by any of these agents, then agent 3 will make a proposal $d^3 = (a^3, t^3) \in \mathcal{D}$ such that: $t_1^3 = t_2^3 = 0$ and d^3 will be submitted to the unanimous approval of agents $\{4, 5, \ldots, n\}$. And so on... If the successive proposals by agents 1, 2, ..., $(n - 2)$ have been successively rejected, then agent $(n - 1)$ proposes $d^{n-1} = (a^{n-1}, t^{n-1}) \in \mathcal{D}$ where t^{n-1} is of the form:

$$t^{n-1} = (0, \cdots, 0, t_{n-1}^{n-1}, t_n^{n-1})$$

Only agent n has the right to veto d^{n-1}, in which case he selects the final decision $d^n = (a^n, 0)$.

1) Prove that the above mechanism defines a dominance-solvable game (up to a certain assumption on tie-breakings, as in Problem 3) and compute the sophisticated equilibrium pay-offs. Is any sophisticated equilibrium a Pareto optimum as well?

2) In our second mechanism a status-quo decision (a^*, t^*) is a priori given. Next the mechanism works by auctioning the

leadership in the following way:

i) each agent bids an amount $\lambda_i \geq 0$ of private good to become the leader;

ii) one agent among the highest bidders (e.g. the agent with the smallest index) becomes the leader - say i_o -;

iii) the leading agent i_o pays λ_{i_o} to every agent. Next he proposes a decision (a; t);

iv) this proposal is submitted to unanimous approval of the other agents. Hence at step iv):

- either agent i_o's proposal (a, t) is accepted by all agents, in which case the final decision is (a, t + s) where

$$s = (\lambda_{i_o}, \ldots, \lambda_{i_o}, \underbrace{-(n-1)\lambda_{i_o}}_{i_o}, \lambda_{i_o}, \ldots, \lambda_{i_o})$$

- or (at least) one agent rejects agent i_o's proposal, in which case the status-quo decision (a^*, t^*) is made. Taking the monetary transfers into account, the final decision is then $(a^*, t^* + s)$.

To analyze the game resulting from this mechanism at some given utility functions $(u_1, \ldots, u_n)$ consider first the game $G(i_o)$ starting at step iii) after the leading agent i_o has been determined.

Prove that $G(i_o)$ is dominance-solvable and compute its sophisticated equilibrium pay-offs. Back to step i) prove that the overall game is not dominance-solvable but that after suitable elimination of dominated strategies, we are left with an inessential game (see Example 5). Compute the corresponding equilibrium pay-offs.

REFERENCES

BERGE, C. 1957. Théorie générale des jeux à n personnes. Paris: Gauthier-Villars.

BINMORE, K. 1980. Nash bargaining theory II, London School of Economics, mimeo.

DUTTA, B. and L. GEVERS. 1981. On voting rules and perfect equilibrium allocation of a shrinking cake.

FARQHARSON, R. 1969. Theory of voting. New Haven: Yale University Press.

GRETLEIN, R. 1980. "Dominance elimination procedures on finite alternative games". School of Urban and Public Affairs, Carnegie Mellon University, Pittsburgh.

MOULIN, H. 1979. "Dominance-solvable voting schemes", Econometrica 47: 1337-1351.

MOULIN, H. 1981. "Implementing just and efficient decision-making", Journal of Public Economics 16.

MOULIN, H. 1981 "The strategy of social choice", Laboratoire d'Econométrie de l'Ecole Polytechnique A229, forthcoming North-Holland Publishing Co.

ROCHET, J.C. 1980. "Selection of a unique equilibrium payoff for extensive games with perfect information", D.P. CEREMADE, Université Paris IX.

RUBINSTEIN, A. 1980. "Perfect equilibrium in a bargaining model", I.C.E.R.D., London School of Economics.

CHAPTER III. NASH EQUILIBRIUM.

Dominant strategy, prudent and sophisticated behaviour can all be sustained independently by the players. Each player on his own, aware only of the normal form of the game can compute the strategy (or strategies) recommended by this or that rationality argument. In particular the timing of various strategical choices plays no role.

On the contrary Nash equilibrium behaviour can be justified only within a dynamic scenario where the strategical decision made today depends on the record of past plays of the game, or at least of yesterday position - so called initial position -. Hence a certain amount of communication across players is now inavoidable, if only through mutual observation of the past outcomes of the game.

We define and discuss the Nash equilibrium concept before going into its mathematical niceties (existence and stability).

I. DEFINITION AND DISCUSSION

Given a N-normal form game, we suppose that the players behave as if they were not aware of their strategic interdependency : when player i considers a switch from strategy x_i to strategy y_i he does not anticipate a reaction to his move by others players i.e. he expects them not to change their strategies in response to his own change. This assumption is plausible if the players are many so that the externality caused by a single deviation to the overall outcome is negligible (see Example 1) ; alternatively in the complete ignorance framework, where player y ignores the utility functions u_j for j in $N \setminus \{i\}$, he can acquire information on the u_j by observing the reactions by $N \setminus \{i\}$ to a switch that would anyway be a profitable one if no reaction occurs (this line of argument is explored in Section 3).

<u>Definition 1.</u>

Given a N-normal form game $G = (X_i, u_i ; i \in N)$, we say that outcome $x = (x_i)_{i \in N}$ is a Nash equilibrium (in short NE) of G if the following holds :

$$\forall\, i \in N \quad \forall\, y_i \in X_i \qquad u_i(y_i, x_{\hat{i}}) \leq u_i(x_i, x_{\hat{i}}) \tag{1}$$

We denote by N E (G) the set of Nash equilibria of G.

At a Nash equilibrium x, player i views the strategies $x_{\hat{i}}$ as exogeneously given, thus maximizing u_i over all conceivable switch y_i. The Nash equilibrium property (1) states that x_i is a best reply strategy to $x_{\hat{i}}$.

The Nash equilibrium concept does not lead to a rationale of behaviour as for sophisticated or prudent behaviour. Typically if a game possesses two non interchangeable N E (see Example 2 and Lemma 2 below) players can not select Nash equilibrium strategies without some coordination mechanism (on this point see also Chapter V Section 2). Notice, however, that for two-person zero sum games, N E outcomes are just saddle-points (Definition 6 Chapter I), so that N E strategies exactly coïncide with prudent-optimal strategies (Theorem 3 Chapter I).

We imagine two extreme scenarios to sustain the Nash equilibrium concept :

1) on the normative side we think of the players openly discussing until they reach a non-binding agreement to play a certain outcome. The next moment they split and all communication between them becomes impossible ; then each player chooses, secretly and ignoring the other actual strategic choice, his actual strategy. He can be faithful to the previous agreement or he can betray it at no cost and use whatever strategy is available to him. If (and only if) the agreed outcome is a Nash equilibrium, then it will be a self-enforcing agreement : assuming that everybody else is loyal, I shall better be loyal myself (and the more conspicuous my virtue, the more incentive I give to the others for being virtuous). This "semi-cooperative" scenario is studied in

Chapter V as a specific cooperative tool : the Nash equilibrium concept is connected both to the non-cooperative and the cooperative side of the theory.

2) on the <u>descriptive</u> side we look for the <u>stable</u> outcomes of a myopic tatonnement process where each player adjusts optimally his strategy under the - constantly violated - assumption that others will not move anymore. When this tatonnement <u>à la Cournot</u> (see Section 3 below) eventually converges, we have reached a Nash equilibrium outcome.

<u>Remark</u> : A meta-theoretical argument proposed by von Neumann and Morgenstern provides an alternative formulation of the normative justification of N E outcomes. Assume complete information and suppose that some theory recommends in game G to each player i the "optimal" strategy x_i. Since each rational and completely informed player can reconstruct the whole theoretical argument, and compute the recommended outcome, it is necessary that our theory suggests a N E outcome if we want selfish utility-maximizing players to actually comply with its command.

<u>Example 1</u>. <u>Binary choices with externalities</u>. (Schelling [1979]).

There are many identical players who must take a binary decision {0,1} (use private automobile or public transportation).

Thus if t , $0 \leq t \leq 1$ is the fraction of players using strategy 1 (public transportation), the numbers $a(t)$ and $b(t)$ denote respectively the utility level of any player using strategy 1 or strategy 0 (private automobile). Hence the following normal form of the game :

$$X_1 = \ldots = X_n = \{0,1\} \quad n \text{ large}$$

$$\left.\begin{aligned} u_i(x_i, x_{\hat{i}}) &= a(t) \text{ if } x_i = 1 \\ &= b(t) \text{ if } x_i = 0 \end{aligned}\right\} \text{ where } t = \frac{1}{n} \sum_{j=1}^{n} x_j$$

Suppose a and b are as in Figure 1

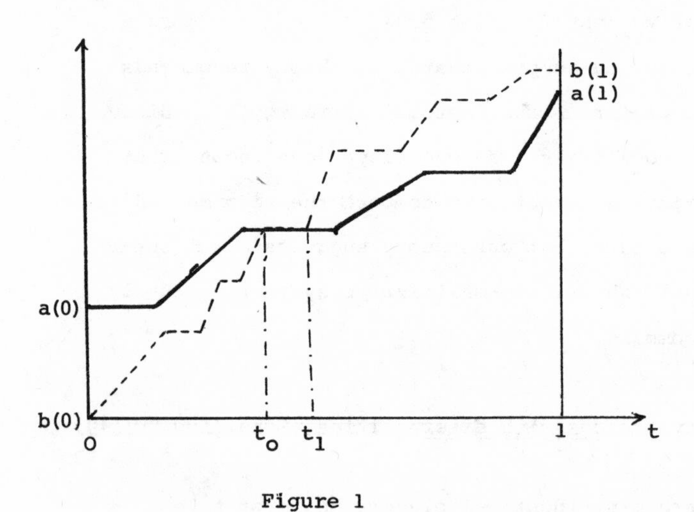

Figure 1

This means that when a fraction above t_1 of the players use public transportation, trafic is so easy that the automobile driver is happier than the bus customer (agregating costs and comfort) whereas if more than $(1-t_o)$ use private automobile, traffic is so bad (and perhaps a bit faster for buses) that the comparison is reversed.

The NE of this game are those outcomes x^* such that

$$t_o \leqslant t^* - \frac{1}{n}, \quad t^* + \frac{1}{n} \leqslant t_1 \quad \text{where} \quad t^* = \frac{1}{n} \sum_{j=1}^{n} x_j^*$$

i.e. such that the marginal player is indifferent between the two available strategies.

Notice that if a fraction δ of the players decide to switch from strategy 0 to 1 and if δ is large enough so that :

$$b(t) = a(t) < a(t+\delta)$$

these players will actually improve upon their utility levels as long as the others do not react. However this move generates an incentive for switching from 1 to 0 as $a(t+\delta) < b(t+\delta)$. This in turn brings back the fraction $\frac{1}{n} \sum_{j=1}^{n} x_j$ down to $[t_o, t_1]$. Similarly if a fraction δ, for obscure reasons (perhaps random mistakes) switch from 1 to 0 and $t - \delta < t_o$, then an incentive to switch back from 0 to 1 emerges as $b(t-\delta) < a(t-\delta)$, thus lifting up the fraction $\frac{1}{n} \sum_{j=1}^{n} x_j$ of 1-players to $[t_o, t_1]$.

Exercise 1.

We consider two different configurations of the pairs $a(.)$, $b(.)$.

1) Suppose a, b are both increasing with t and :

$$a(t) < b(t) \quad \text{all} \quad t \in [0,1]$$

Prove that the corresponding game is a generalized Prisonner's Dilemna (Example 1 Chapter I) where the unique N E is a dominating strategy equilibrium as well.

2) Suppose a, b are as an Figure 2. Prove that the corresponding game has three different types of N E out of which one is unstable with respect to δ - perturbations as above.

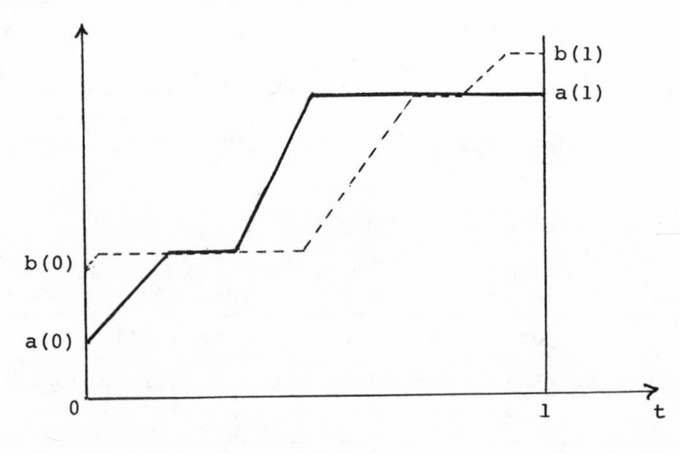

Figure 2

We postpone until Section 3 the formal analysis of the tatonnement à la Cournot, a crucial step along the non cooperative interpretation of the Nash equilibrium.

Lemma 1.

An outcome x of game $G = (X_i, u_i ; i \in N)$ is said to be individually rational if we have :

$$\sup_{y_i \in X_i} \inf_{y_{\hat{i}} \in X_{\hat{i}}} u_i(y_i, y_{\hat{i}}) = \alpha_i \leqslant u_i(x) \quad \text{all } i \in N \qquad (2)$$

All N E outcomes are individually rational.

Proof.

From (1) we get for all i :

$$\forall y_i \in X_i \quad \inf_{y_{\hat{i}} \in X_{\hat{i}}} u_i(y_i, y_{\hat{i}}) \leqslant u_i(y_i, x_{\hat{i}}) \leqslant u_i(x)$$

Taking the supremum over y_i of these inequalities yields (2).

■

On one hand a N E outcome provides each player with at least his secure utility level (individual rationality), although a N E strategy may fail to be a prudent one (see Example 2). On the other hand a N E outcome can be Pareto dominated (as in Example 1 above). Moreover even if every N E outcome is Pareto optimal, coexistence of several distinct Pareto optimal N E outcomes raises a typical struggle for the leadership (Chapter

II, Section 4) thus killing any hope of finding out "optimal" strategies. This point is illustrated by our next example.

Example 2. The crossing game.

Our players drive on two orthogonal roads and reach the crossing simultaneously.

Each player can stop or go. The following 2 x 2 game

	stop	go
stop	1, 1	$1-\varepsilon$, 2
go	2, $1-\varepsilon$	0, 0

formalizes the situation by assuming that each player prefers to stop than suffering an accident (outcome (go, go)) and each prefers to go if the other stops. The non negative number ε is the utility loss that results from viewing the other passing while one is gently stopping ; it varies according to the cultural pattern.

The two N E outcomes (namely (stop, go) and (go,stop)) are Pareto optimums as well. However they are not interchangeable : it is optimal for me to stop if you go, but to go if you stop. By committing himself to the non prudent strategy go, a player wins, since he forces the other to stop and therefore enjoys the maximal utility level 2. Since no outcome offers the utility level 2 to both players, the struggle for the leadership emerges. Each player will pretend that he has lost his ability to switch from go to stop (e.g. by looking drunk) while at the same time he furiously observes the opponent to check whether or not he really can not stop anymore. The striking feature of these tactical moves is that a skillful irrationality in one's behaviour is winning, a quite rational attitude after all. The symmetry of both players' roles makes it impossible to arbitrate the struggle for the leadership by means of the sole normal form of the game. For superb examples and deep comments we refer the reader to Schelling [1971] Chapter II.

The conclusions of the previous example are easily extended to a general two-person game $G = (X_1, X_2, u_1, u_2)$. Using the notations of Definition 5 Chapter II, we call i - Stackelberg utility level and we denote by S_i player i's utility at any i-Stackelberg equilibrium :

$$S_i = \sup_{(x_1,x_2) \in BR_j} u_i(x_1,x_2) \qquad \text{all } \{i,j\} = \{1,2\}$$

Thus S_i is the utility level of player i acting optimally as a leader. We say that the struggle for the leadership happens in G if there is no outcome x such that :

$$S_i \leq u_i(x) \qquad i = 1, 2 \tag{3}$$

Lemma 2.

Suppose G has at least two Pareto optimal N E outcomes x^1, x^2 with distinct associated utility vectors :

$$(u_1(x^1), u_2(x^1)) \neq (u_1(x^2), u_2(x^2)) \qquad (4)$$

Then the struggle for the leadership happens in G.

Proof.

Observe that $NE(G) = BR_1 \cap BR_2$. Therefore by definition of S_i we have :

$$\{x \in NE(G)\} \Rightarrow \{u_i(x) \leq S_i \quad i = 1, 2\}$$

If the struggle for the leadership does not happen in G, there is an outcome x such that (3) holds, implying :

$$u_i(x^1) \leq u_i(x) \qquad i = 1, 2$$

$$u_i(x^2) \leq u_i(x) \qquad i = 1, 2$$

Since x^1 and x^2 both are Pareto optimums, these four inequalities are bound to be equalities thus violating assumption (4).

■

Our next step in the discussion of the N E outcomes is to compare them with the sophisticated equilibrium outcomes.

Theorem 1.

Suppose that X_i is finite for all i.

If the game $G = (X_i, u_i ; i \; N)$ is dominance-solvable its sophisticated equilibrium outcomes all are Nash equilibrium outcomes.

Proof.

The proof parallels that of Lemma 4 Chapter II. One shows that for any rectangular subset $Y = \underset{i \in N}{X} Y_i$ of X_N we have :

$$N E (Z) \subset N E (Y) \quad \text{where} \quad Z_i = \mathcal{D}_i (u_i ; Y)$$

where $N E(Y)$ stands for the set of $N E$ outcomes of game $(Y_i, u_i ; i \in N)$.

Next if $X^t = \underset{i \in N}{X} X_i^t$ denotes the set of sophisticated equilibria of G we have :

$$N E (X) \supset N E (X^1) \supset \ldots \supset N E (X^t) = X^t$$

■

Thus for dominance-solvable games, sophisticated behaviour always leads to a N E outcome. The converse statement is far from being true : <u>a N E strategy might very well be a dominated strategy</u>! Let us illustrate this striking fact on a 3 x 3 game (i.e. a two person normal form game where each player has 3 strategies).

	L		R
T	1, 1	0, 1	0, 1
	0, 0	1, 0	0, 1
B	1, 0	0, 1	1, 0

Here (Top, Left) is the unique N E outcome. However Top is dominated by Bottom for player 1 (row) and Left is dominated by Right for player 2 (column). After elimination of dominated strategies the players face a zero sum game without saddle-pair.

The implications of this remark are more transparent on extensive form games.

Example 3. Voting by veto. (continued)

From the 3 player game of Example 2 Chapter II, we extract a 2-player game by assuming that player 1 vetoes outcome b (i.e. his sophisticated behaviour in view of the given preferences). Thus player 2 and 3 face the following game :

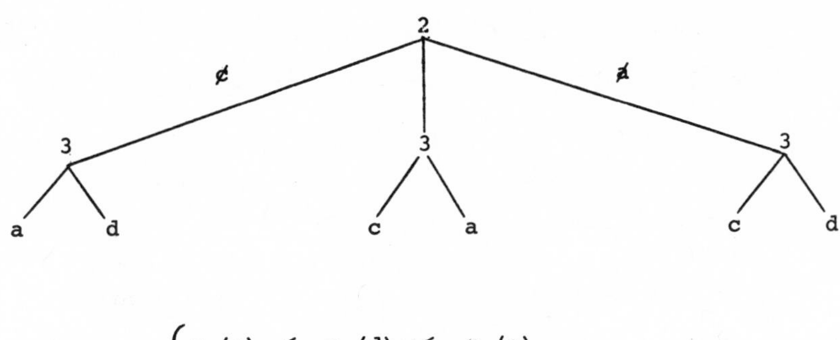

$$\begin{cases} u_2(c) < u_2(d) < u_2(a) \\ u_3(c) < u_3(a) < u_3(d) \end{cases}$$

Figure 3

Three N E outcomes emerge : ($\not d$, R) ($\not c$, L) and ($\not a$, L) .

α) In ($\not d$, R) player 2 vetoes d and player 3 moves always right : R is player 3's dominating strategy and ($\not d$, R) the sophisticated equilibrium outcome. Candidate a is elected.

β) In ($\not c$, L) and ($\not d$, L) player 2 vetoes respectively c or d whereas player 3 uses the following strategy L :

$$\begin{cases} \text{move right if player 2 vetoes c or a} \\ \text{move \underline{left} if player 2 vetoes d} \end{cases}$$

The best reply strategies of player 2 to L are precisely

$\{\not c, \not d\}$. Moreover L is a best reply to both $\not c$ and $\not d$ since the corresponding elected outcome is d, the top candidate of u_3. Thus $(\not c, L)$ and $(\not d, L)$ are both N E outcomes.

Notice that L not only is a dominated strategy (by R) for player 3 but also a risky (i.e. non prudent) one : by L, player 3 threatens to elect c if player 2 ever vetoes d, but carrying out this threat is harmful to both players (c is the bottom candidate of u_2 and u_3).

Another example of N E outcome involving dominated strategies is our simple auction model : see Exercize 2 below.

Notice, however, that when all utility functions are one-to-one on X_N, no N E outcome contains a dominated strategy and the N E concept does generalize D S E.

Exercise 2.

In the second price auction (Example 2 Chapter I) prove that for all player i and all price p bounded above by the value of the object to i ($p \leq a_i$), there exists a N E outcome where i gets the object at price p.

On the contrary in the first price auction show that any N E outcome is such that player 1 gets the object at a price between a_2 and a_1.

Problem 1. *Almost inessential two-person game.*

Let $G = (X_1, X_2, u_1, u_2)$ be a two person game where X_1, X_2 are both finite. We say that G is almost inessential if all individually rational outcomes have the same pay-off vector. Thus for any two outcomes x, y :

$$\{x \text{ and } y \text{ are individually rational}\} \Rightarrow \{u_i(x) = u_i(y),\ i = 1,2\}$$

1) Give an example of an almost inessential game which is not inessential.

2) Suppose that G is almost inessential and pick a pair $(x_1, x_2) \in P_1(u_1) \times P_2(u_2)$ of prudent strategies. Show that outcome (x_1, x_2) is a Pareto optimum, a N E, an i-Stackelberg equilibrium for $i = 1, 2$.

3) Is any N E in an almost inessential game an i-Stackelberg equilibrium as well (for $i = 1, 2$)? What if both utility functions are one-to-one on $X_1 \times X_2$?

II. EXISTENCE OF THE NASH'S THEOREM

From a theoretical point of view, the most appealing feature of the Nash equilibrium concept is its nice mathematical tractability.

Nash's theorem provide sufficient conditions for a game to possess at least one Nash equilibrium : these conditions prove to be easily applicable in many applied models.

We have already discovered two distinct sufficient conditions for a game G to possess at least one N E outcome:

i) if G is inessential : then by Theorem 1 Chapter I, all N-uples of prudent optimal strategies are N E as well.

ii) if G is dominance-solvable : Theorem 1 of this Chapter.

However no useful condition exists on X_i and u_i, guaranteing inessentiality or dominance-solvability of G. The typical situation is that of a normal form game where u_i are built up from elementary functions (polynomials, log.,...) by elementary operations. In that framework Nash's result proves to be quite helpful.

Theorem 2. (Nash [1951]).

Suppose that for all $i \in N$, the strategy sets X_i are convex and compact subsets of some topological vector spaces (that can vary with i). Suppose that for all $i \in N$, u_i is a continuous real valued function defined on X_N

such that :

for all $x_{\hat{i}} \in X_{\hat{i}}$ $u_i(x_i, x_{\hat{i}})$ is a concave function of x_i on X_i

Then the set $N E (G)$ of the Nash equilibrium outcomes of the game $G = (X_i, u_i, i \in N)$ is non empty and compact. Recall that a real valued function α is concave if for all λ, $0 \leqslant \lambda \leqslant 1$, and all t, t' we have :

$$\lambda \alpha (t) + (1 - \lambda) \alpha (t') \leqslant \alpha (\lambda t + (1 - \lambda) t')$$

Proof.

The proof requires the use of a fixed point-type argument from convex analysis. We shall use the following result known as the Knaster - Kuratowski - Mazurkewitch Lemma (a proof of which can be found in Berge [1957]) .

Lemma.

Let p be any integer and $a_1, \ldots, a_p$ be some points in a topological vector space. Let further $A_1, \ldots, A_p$ be some closed subsets of $C O \{a_1, \ldots a_p\}$, the convex hull of $\{a_1, \ldots, a_p\}$ such that :

$$\forall T \subset \{1, \ldots, n\} \quad \bigcup_{k \in T} A_k \text{ contains } C O \{e_k / k \in T\}$$

Then the intersection $\bigcap_{k=1}^{p} A_k$ is non empty.

The following short proof of Theorem 2 is due to Georges Haddad.

Let us define the real valued function ϕ on $X_N \times X_N$:

$$\phi(x,y) = \sum_{i \in N} [u_i(x_i, y_{\hat{i}}) - u_i(y)] \quad \text{all } x, y \in X_N$$

It follows from the concavity of u_i w.r.t.x_i that ϕ is concave w.r.t.x. In addition ϕ is continuous w.r.t.y. We define now a multivalued function F from X_N into itself :

$$F(x) = \{ y \in X_N \,/\, \phi(x,y) \leq 0 \} \quad \text{all } x \in X_N$$

Since ϕ is continuous in y, F(x) is compact for all x. Moreover $x \in F(x)$ so that F is non empty valued. We fix an integer p and p elements $x^1, \ldots, x^p$ in X_N. We claim that any convex combination $x = \sum_{k=1}^{p} \lambda_k x^k$ belongs to $\bigcup_{k=1}^{p} F(x^k)$. For if it would not we would have :

$$\forall k = 1, \ldots, p \qquad 0 < \phi(x^k, x)$$

By the concavity of ϕ w.r.t. its first variable, this yields a contradiction :

$$0 < \phi\left(\sum_{k=1}^{p} \lambda_k x^k, x\right) = \phi(x, x) = 0$$

Thus we have proved :

$$C\,O\,[x^1, \ldots, x^p] \subset \bigcup_{k=1}^{p} F(x^k)$$

Because this holds for all p and all $x^1, \ldots, x^p$ we deduce from the K K M Lemma that $\bigcap_{k=1}^{p} F(x^k)$ is non-empty. Hence the non empty compact sets $(F(x))_{x \in X_N}$ are such that any finite subfamily of them has non-empty intersection. Thus the overall intersection $\bigcap_{x \in X_N} F(x)$ is non empty. Any x^* in this intersection is such that :

$$\forall x \in X_N \quad \phi(x, x^*) \leq 0$$

which is easily rewritten as :

$$\forall i \in N \quad \forall x_i \in X_i \quad u_i(x_i, x_{\hat{i}}^*) - u_i(x^*) \leq 0$$

Thus $\bigcap_{X_N} F(x) = NE(G)$ and the proof of Theorem 2 is complete.

■

<u>Corollary of Nash's Theorem</u> (Von Neumann).

Let X_1, X_2 be two convex compact subsets of some topological vector spaces and let u_1 be a continuous real valued function defined on $X_1 \times X_2$ such that :

- for all $x_2 \in X_2$, $u_1(x_1, x_2)$ is concave w.r.t. x_1
- for all $x_1 \in X_1$, $u_1(x_1, x_2)$ is convex w.r.t. x_2.

Then the two-person zero sum game (X_1, X_2, u_1) has at least a saddle pair (and therefore a value).

Nash's theorem allows us to claim that the set $NE(G)$ is non empty. To actually compute it requires to solve simultaneously the following equations :

$$u_i(x^*) = \max_{x_i \in X_i} u_i(x_i, x^*_{\hat{i}}) \tag{5}$$

If u_i is concave with respect to x_i the above global maximization problem is equivalent to a local problem (as we know from convex programming). For instance if x_i is interior to X_i and u_i is differentiable w.r.t.x_i conditions (5) are equivalent to :

$$\frac{\partial u_i}{\partial x_i}(x^*) = 0 \qquad \text{all } i \in N \tag{6}$$

System (6) is expected to have isolated solutions as its number of independent equations equals the dimension of X_N. This implies for instance that <u>Nash equilibrium outcomes are not, in general, Pareto optimal</u> (for a precise genericity result, see Grote [1974]) .

Let us illustrate the method that we have roughly described on some examples and problems.

<u>Example 4</u>. <u>Quantity setting oligopoly</u>.

A finite number n of costless producers control the supply $x_1, \ldots, x_n$ of some satiable good that they offer on the market of this good. If the overall supply is

$\overline{x} = x_1 + \ldots + x_n$, the price becomes $p(\overline{x})$ where p is decreasing and concave on the positive half line :

$$p(0) > 0 \quad p'(y) < 0 \quad p''(y) < 0 \ , \quad \text{all } y \geq 0 \qquad (7)$$

The situation is summarized in the following game :

$$X_i = [\,0,+\infty[\qquad u_i(x) = x_i \cdot p(\overline{x}) \quad \text{all } i = 1,\ldots,n$$

From our assumptions on p we deduce that u_i is concave with respect to its variable x_i. As X_i are not compact, we set $Y_i = [\,0,S\,]$ where S is the supply at which the price vanishes : $p\,(S) = 0$. To the restricted game $(Y_i\,, u_i;\ i = 1,\ldots,n)$ Nash's theorem applies, therefore proving the existence of a N E outcome x for the restricted game. Actually x is a N E of the original game. Namely the secure utility level of any player in the restricted game is 0 , hence, by Lemma 1 :

$$u_i(x) \geq 0 \geq u_i(y_i, x_{\hat{i}}) \quad \text{for all} \quad y_i \in [\,S, +\infty[$$

By the concavity and differentiability of u_i over X a N E outcome x is characterized by system (6) :

$$x_i \cdot p'(\overline{x}) + p(\overline{x}) = 0$$

Thus our game has a unique N E outcome on the diagonal :

$$x_1 = \ldots = x_n = \tilde{x} \quad \text{where} \quad \tilde{x} = -\frac{p}{p'}(n\,\tilde{x})$$

Exercise 3. Example 4 (continued)

Prove that the overall supply $n\,\tilde{x}$ at the N E outcome strictly exceeds the supply $x^\star$ that maximizes the joint profit $\bar{x}\,p\,(\bar{x})$.

Prove that any strategy x_i such that $x^\star < x_i$ is dominated by strategy $x^\star$. Is our game dominance solvable ?

Exercise 4. Quantity setting duopoly.

We assume that the price varies now as :

$$p\,(\bar{x}) = [\frac{1}{\bar{x}} - 1]^{\frac{1}{2}}$$

and consider the duopoly (two players) game where each player's strategy set is $[\,0, \frac{1}{2}\,]$. Prove that $u_i(x) = x_i \,.\, p(\bar{x})$ is concave w.r.t. x_i. Compute the best reply function of both players, and the N E outcome. Compare it with the Pareto optimums line.

Problem 2. The auto-dealer game, (Case [1980]).

The n players are n auto-dealers facing a constant overall fixed demand D. Let the strategy x_i be the number of cars that dealer i keeps on hand. Assuming that each dealer has the same number of visitors per unit of time, dealer i faces the demand-flow :

$$D.\ \frac{x_i}{\bar{x}} \quad \text{where } \bar{x} = x_1 + \ldots + x_n$$

Let P_i be his unit profit and C_i his unit cost of storing (per unit of time). Then the following n-normal form game emerges :

$$\begin{cases} X_i = [0,+\infty[\quad u_i(x) = D\, P_i \frac{x_i}{\bar{x}} - C_i\, x_i, \quad \text{all } i = 1,\ldots,n \\ \text{with the convention } \frac{0}{0} = 1. \end{cases}$$

1) Compute the undominated strategies of player i. Is our game dominance-solvable ?

2) Prove that our game has two NE outcomes and compute them. Are they Pareto optimal outcomes ?

3) Compare the relative NE profits when the dealers split in two subsets $\{1,\ldots,n\}$ $\{n_1+1,\ldots,n\}$ such that

for $i = 1,\ldots, n_1$: $\frac{C_i}{P_i}$ is negligible

for $i = n_1+1,\ldots,n$: $\frac{C_i}{P_i}$ is constant, strictly positive.

Problem 3. Provision of a public good by subscription.

In an economy with one public good (of which the available quantity is denoted q) and one private good (of which the endowment of player i is denoted m_i) we write $v_i(q, m_i)$ the utility function of player i, and we assume that v_i is differentiable, strictly monotonic in each variable and concave as a function of (q, m_i). We assume that the public good is produced from the private good on a one-to-one basis. Let M_i be the initial endowment of money of player i .

Each player voluntarily prescribes an amount x_i of his private good resources to contribute to the overall production of public good. Thus we face the following N - normal form game :

$$X_i = [0, M_i], \quad u_i(x) = v_i(\sum_{j \in N} x_j, M_i - x_i)$$

1) Prove that outcome x is Pareto optimum within this game if and only if it satisfies the Lindahl-Samuelson equation :

$$\sum_{j \in N} \left(\frac{\partial v_i}{\partial q} \Big/ \frac{\partial v_i}{\partial m_i} \right) (x) = 1$$

2) Prove that u_i is concave w.r.t. its variable x_i and deduce the existence of a NE outcome. Characterize the NE outcomes by $|N|$ first order conditions.

3) Prove that at any NE outcome production of the public good is suboptimal. More precisely from a NE outcome, a Pareto improving move exists where the production of public good increases.

Problem 4. *A two person zero-sum game.*

Let A_1, A_2 be two definite positive $p \times p$ matrices, let B be any $p \times p$ matrix and finally let $a_1, a_2 \in \mathbb{R}^p$.

Consider the two person zero-sum game (X_1, X_2, u_1) where :

$$X_1 = X_2 = \mathbb{R}^p$$

$$u_1(x_1,x_2) = -\frac{1}{2} < A_1x_1, x_1 > + < Bx_1,x_2 > + \frac{1}{2} < A_2x_2,x_2 > + < a_1,x_1 > + < a_2,x_2 >$$

Prove that this game has a unique saddle pair and compute it.

Problem 5. *The war of attrition, (Milgrom Weber [1980]).*

Two players compete for a single indivisible object. This object is worth v_i to player i. The winner of the object is this player who remains agressive longer, given that the cost of being agressive is one per unit of time. Thus player i's strategy x_i means : I shall remain agressive until time $t = x_i$ unless the other quits at some time $x_j < x_i$; in this latter case I quit at time $x_j + \varepsilon$ thus winning the object at that cost.

1) The following normal form game is in order :

$$X_1 = X_2 = [0,+\infty[$$

$$u_1(x_1,x_2) = v_1 - x_2 \text{ if } x_2 < x_1 \qquad u_2(x_1,x_2) = -x_2 \quad \text{if } x_2 < x_1$$

$$= -x_1 \quad \text{if } x_1 < x_2 \qquad\qquad = v_2 - x_1 \text{ if } x_1 < x_2$$

$$= \frac{v_1}{2} - x_1 \text{ if } x_1 = x_2 \qquad\qquad = \frac{v_2}{2} - x_2 \text{ if } x_1 = x_2$$

(ties are broken by flipping a coin to allocate the object).

Prove that our game has exactly three Pareto optimal pay-off vectors, two of which are also Nash equilibrium pay-off vectors (hence a struggle for the leadership).

Prove that in any N E outcome, one player is using his unique prudent strategy (and gets his secure utility level) whereas the other uses a very risky (i.e. non prudent) strategy.

2) Suppose now that v_1, v_2 are two independent random variables with uniform distribution on $[0, 1]$. Player i observes v_i but not v_j. His strategy set is now $\tilde{X}_i$:

$$\tilde{x}_i \in \tilde{X}_i : \tilde{x}_i \text{ is a maesurable mapping from } [0,1] \text{ into } X_i$$

And the pay-off functions become :

$$\tilde{u}_i(\tilde{x}_1, \tilde{x}_2) = \int_{[0,1]^2} u_i(\tilde{x}_1(v_1), \tilde{x}_2(v_2))\, d\, v_1 \;\; d\, v_2$$

Prove the existence in game $(\tilde{X}_1, \tilde{X}_2, \tilde{u}_1, \tilde{u}_2)$ of a symmetrical, Pareto dominated, N E outcome : $\tilde{x}_1 = \tilde{x}_2 = x^\star$, and compute it. Observe that wars of arbitrary large length do occur at this equilibrium, but the expected length of the war is finite.

Hint : Assume that $x^\star$ is increasing and differentiable. Then prove that for all $v_i \in [0,1]$ the function :

$$\varphi(y) = \int_0^{(x^\star)^{-1}(y)} (v_i - x^\star(t))\, dt - y\,[1-(x^\star)^{-1}(y)]$$

must be maximal on $[0, +\infty[$ at $y = x^\star(v_i)$.

III. STABLE EQUILIBRIA

Even under the complete information assumption (each player knows the entire normal form of the game, including other players' pay-offs) the Nash equilibrium concept can not be justified by decentralized rationality arguments. If communication is feasible one can invoke the non-binding agreement scenario (see Section 1). On the contrary, the Cournot tatonnement process explores the dynamic consequences of the myopic, perfect competition-like, behavioural assumption that each player maximizes his or her pay-off by taking the other strategies as fixed. Although it can not be justified by rationality arguments (since the myopic assumption that the other players will keep their current strategies is constantly violated) it has a transparent descriptive power and allows us to partition stable / unstable NE outcomes. Moreover it requires only the minimal informational assumption that we call "complete ignorance" (each player is aware of his own utility only and communication is feasible only via mutual observation of strategies).

Example 5. Stability in a Cournot quantity setting duopoly.

The two players supply respectively the quantities x_1 and x_2 of the same commodity, for which the price is then settled as

$$p(\bar{x}) = 1 - \bar{x} \quad \text{where} \quad \bar{x} = x_1 + x_2$$

We shall consider two distinct assumptions on the cost function :

α) - constant returns to scale (C RS) : the cost of producing y units of the commodity is $\frac{1}{2}y$ for both players.

β) - increasing returns to scale (I RS) : the cost of producing y units is $\frac{1}{2}y - \frac{3}{4}y^2$ for both players.

Finally each player maximal production capacity is $\frac{1}{2}$ (so that p and the costs are never negative).

The C R S - game is then :

$$X_1 = X_2 = [0, \frac{1}{2}] \quad u_i(x_1, x_2) = x_i(1-\bar{x}) - \frac{1}{2} x_i, \quad i = 1,2$$

The best reply of player i to strategy x_j of player j is easily computed (as u_i is concave w.r.t. x_i).

$$BR_i = \{x_i = \alpha(x_j) / 0 \leq x_j \leq \frac{1}{2}\} \text{ where } \alpha(y) = \frac{1}{4} - \frac{1}{2} y.$$

(we use the notation introduced in Section 4 Chapter II).

The unique N E outcome is :

$$NE = BR_1 \cap BR_2 = \{(\frac{1}{6}, \frac{1}{6})\}$$

Starting from an outcome (x_1^o, x_2^o) the Cournot tatonnement goes by each player successively picking a best reply strategy to the current strategy of the opponent:

$$(x_1^o, x_2^o) \to (x_1^1, x_2^o) = (\alpha(x_2^o), x_2^o) \in BR_1 \to (x_1^1, x_2^1) = (x_1^1, \alpha(x_1^1)) \in BR_2 \to$$
$$\dots \to (x_1^t, x_2^{t-1}) = (\alpha(x_2^{t-1}), x_2^{t-1}) \in BR_1 \to (x_1^t, x_2^t) = (x_1^t, \alpha(x_1^t)) \in BR_2 \to \dots \tag{8}$$

On figure 4 we have drawn two such sequences ; a few more trials will convince the reader that from any starting point (x_1^o, x_2^o) the sequences (x_1^t, x_2^t) as well as (x_1^t, x_2^{t-1}) converge to the N E outcome $(\frac{1}{6}, \frac{1}{6})$. (We leave as an exercize to prove this claim, and that the convergence occurs at a geometric rate). Thus we say that $(\frac{1}{6},\frac{1}{6})$ is a <u>stable</u> N E outcome.

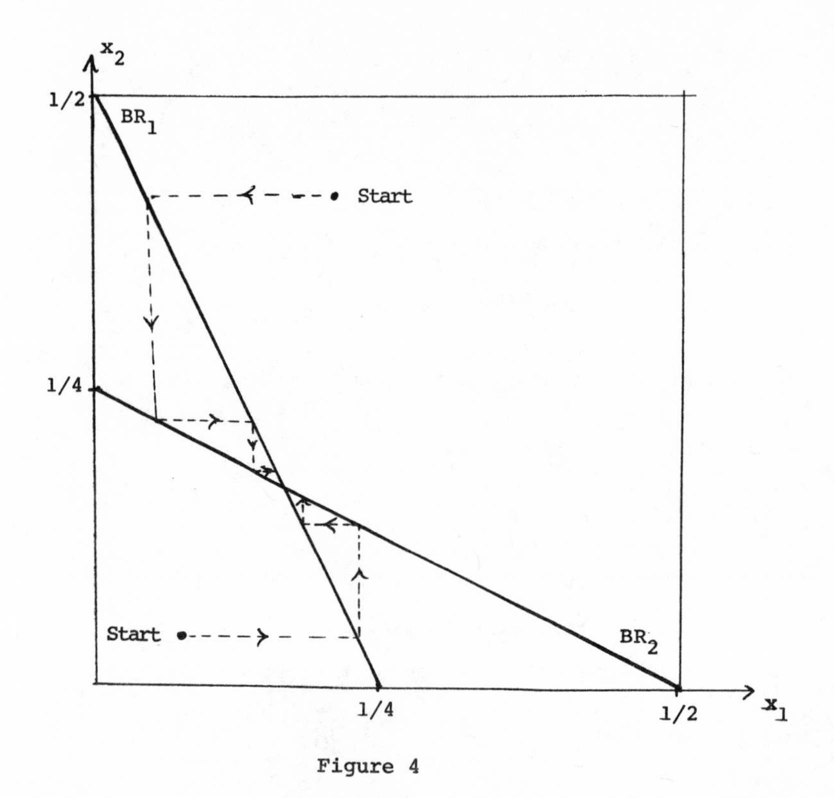

Figure 4

We turn now to the IRS-game, of which the normal form is :

$$X_1 = X_2 = [0,1], \; u_i(x_1,x_2) = x_i(1-\bar{x}) - (\frac{1}{2}x_i - \frac{3}{4}x_i^2), \; i=1,2$$

Again u_i is concave w.r.t. x_i and the best reply curves are :

$$B R_i = \{x_i = \beta(x_j) \,/\, 0 \leq x_j \leq \tfrac{1}{2}\} \text{ where } \beta(y) = \tfrac{1}{2} \text{ if } 0 \leq y \leq \tfrac{1}{4}$$

$$= 1-2y \text{ if } \tfrac{1}{4} \leq y \leq \tfrac{1}{2}$$

Now we have three N E outcomes :

$$N E = B R_1 \cap B R_2 = \{(\tfrac{1}{3}, \tfrac{1}{3}), (\tfrac{1}{2}, 0), (0, \tfrac{1}{2})\}$$

On Figure 5 we observe that starting from any outcome $x^o \neq (\frac{1}{3}, \frac{1}{3})$ the sequence (8) always converges toward $(\frac{1}{2}, 0)$ or $(0, \frac{1}{2})$ (actually in finitely many steps). This holds even when x^o is arbitrarily close to, but different from $(\frac{1}{3}, \frac{1}{3})$. We shall say that $(\frac{1}{3}, \frac{1}{3})$ is an unstable N E outcome whereas $(\frac{1}{2}, 0)$ and $(0, \frac{1}{2})$ are both (locally) stable.

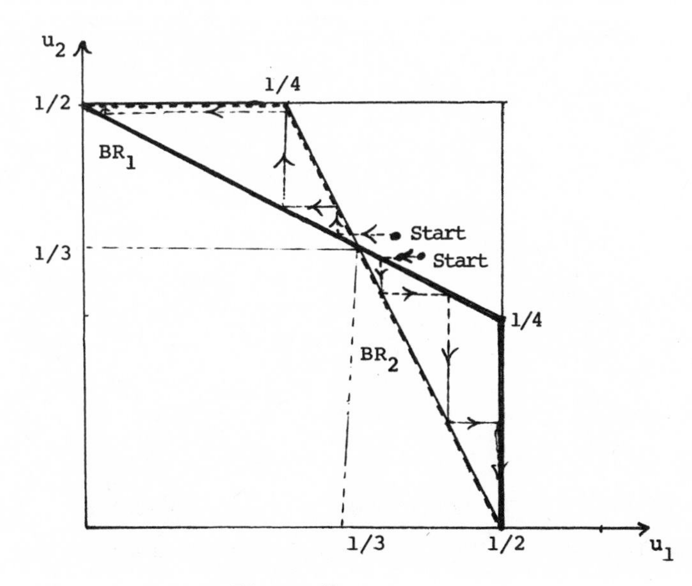

Figure 5

Exercise 5.

On Figure 6 is represented a possible configuration of the best reply curves BR_i, $i = 1,2$ for a game on the strategy sets $X_1 = X_2 = [0,1]$. Check that neither of the 3 N E outcomes is stable. The - very complex - structure of the dynamic system (8) is analyzed for a similar game by Rand [1978] .

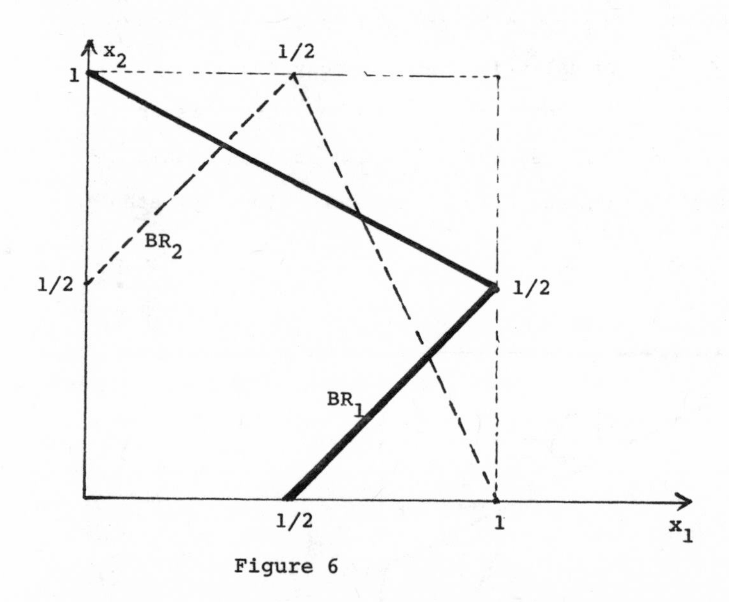

Figure 6

In n-person games the Cournot-tatonnement can be given several definitions : the players can adjust their strategies successively (in which case their ordering does matter) or simultaneously. The corresponding stability notions coïncide for two player games (n=2) but not for games with at least three players ($n \geqslant 3$) : see Exercize 6 below. We adopt the simultaneous definition :

<u>Definition 2.</u>

Let X_i be endowed with some topology for all $i \in N$. Let $G = (X_i, u_i ; i \in N)$ be a N-normal form game. We assume that every player has a unique best reply strategy to any fixed strategies by the other agents :

$$\text{For all } i \in N \text{ and all } x_{\hat{\imath}} \in X_{\hat{\imath}} \text{ there is a unique } r_i(x_{\hat{\imath}}) \in X_i \text{ such that } (r_i(x_{\hat{\imath}}), x_{\hat{\imath}}) \in BR_i \tag{9}$$

To any outcome $x^o \in X_N$ we associate the (simultaneous) Cournot-tatonnement starting at x^o, namely the following sequence $x^o, x^1, \ldots, x^t, \ldots$ of X_N :

$$x_i^t = r_i(x_{\hat{\imath}}^{t-1}) \quad \text{all } i \in N, \quad \text{all } t = 1, 2 \ldots \tag{10}$$

We say that <u>a Nash equilibrium outcome x^* is stable in</u> G if for any initial position $x^o \in X_N$ the Cournot-tatonnement starting at x^o converges to x^*.

Notice that a stable N E outcome of G necessarily is the <u>unique</u> NE of G, for if the initial position is a N E, the Cournot-tatonnement is a constant sequence.

Exercise 6. Stability by sequential Cournot-tatonnement.

Given an ordering $N = \{1,2,\ldots,n\}$ of the players, the $\{1,\ldots,n\}$ - sequential Cournot-tatonnement starting at x^o is the sequence $x^o, x^1, \ldots, x^t, \ldots$ where :

$$x_i^t = r_i(x_1^t, \ldots, x_{i-1}^t, x_{i+1}^{t-1}, \ldots, x_n^{t-1}) \text{ all } i \in N, \text{ all } t = 1,2\ldots$$

Next we say that a N E outcome x^* is $\{1,\ldots,n\}$-stable if for any initial position x^o the $\{1,\ldots,n\}$-sequential Cournot-tatonnement starting at x^o converges to x^*.

1) If $n = 2$, the $\{1, 2\}$-stability, the $\{2,1\}$-stability and the stability proposed by Definition 2 all coïncide.

2) If $n \geq 3$ these notions differ. Consider for instance the 3 players game :

$$X_i = \mathbb{R}, \ i = 1,2,3 \begin{cases} u_1(x) = -(x_1 - x_2)^2 \\ u_2(x) = -(x_2 - \frac{1}{3} x_3)^2 \\ u_3(x) = -(4 x_1 - 3 x_2 - x_3)^2 \end{cases}$$

The unique N E outcome is $(0,0,0)$. Prove that it is not $\{1,2,3\}$-stable whereas it is $\{2,1,3\}$-stable. Is it stable in the sense of Definition 2 ?

Sufficient conditions for a N E outcome to be stable are difficult to obtain and turn out to be quite restrictive (see Gabay-Moulin [1980], Okuguchi [1976], Rosen [1965]).

However if we weaken the stability condition by requiring only that the Cournot-tatonnement starting <u>nearly</u> x^* should in the limit approach x^* we are able to characterize almost completely the (locally) stable N E outcomes.

<u>Definition 3</u>. (same notation as in Definition 1.)

We say that a Nash equilibrium outcome x^* is <u>locally stable</u> in G if there exists for all $i \in N$ a neighborhood V_i of x_i such that assumption (9) holds on V_N and x^* is stable in the restricted game $(V_i, u_i ; i \in N)$.

To derive a computational characterization of local stability, we assume that for all $i \in N$, X_i is a subset of an euclidian space E_i and we fix a N E x^* such that x_i^* is an interior point of X_i, all $i \in N$.

We assume moreover that the utility functions u_i are twice continuously differentiable in a neighbourhood of x_i and that the second derivative $\frac{\partial^2 u_i}{\partial x_i^2}$ is a definite-negative operator at x^*. (therefore (9) holds in a suitable neighbourhood of x^*).

We define a linear operator T from $E_N = \underset{i \in N}{X} E_i$ into itself :

$$\text{for all } e \in E_N : T_i(e) \sum_{j \in N \setminus \{i\}} \left(\frac{\partial^2 u_i}{\partial x_i^2}\right)^{-1} \left(\frac{\partial^2 u_i}{\partial x_i \partial x_j}\right) (e_j) \quad (11)$$

where all the above derivatives are taken at $x^\star$.

Theorem 3.

Suppose that the modulus of all eigenvalues of T is strictly less than 1. Then $x^\star$ is a locally stable N E outcome.

Suppose that $x^\star$ is a locally stable N E outcome. Then the modulus of all eigenvalues of T is less than or equal to 1.

Proof.

Since $\frac{\partial u_i}{\partial x_i}$ is C^1 -differentiable, and $\frac{\partial^2 u_i}{\partial x_i^2}$ is non singular at $x^\star$, the implicit function theorem shows that r_i is, locally at $x^\star$, a C^1 -differentiable function from $X_{N \setminus \{i\}}$ into X_i. Henceforth system (10) can be - locally - written as :

$$x^t = f(x^{t-1}) \quad (12)$$

where f is C^1- differentiable from E_N into itself.

By assumption x^* is a fixed point of f. Thus, by an elementary result in dynamic systems (see e.g. Ortega Rheinboldt(1970) we know that x^* is a locally stable solution of (12) if the modulus of all eigenvalues of $f'(x^*)$ is strictly less than 1. Conversely if at least one eigenvalue of $f'(x^*)$ has a modulus strictly above 1, then x^* is not locally stable. We let the reader check that $T = f'(x^*)$.

■

Corollary.

Suppose n = 2 and X_1, X_2 are one dimensional. Let x^* be a N E of $G = (X_1, X_2, u_1, u_2)$ such that :

- x_i^* is an interior point of X_i
- u_i is C^2-differentiable in a neighbourhood of x^*
- $\frac{\partial^2 u_i}{\partial x_i^2}(x^*) < 0$

Then we have :

$$\left| \frac{\partial^2 u_1}{\partial x_1 \partial x_2} \cdot \frac{\partial^2 u_2}{\partial x_1 \partial x_2} \right| < \left| \frac{\partial^2 u_1}{\partial x_1^2} \cdot \frac{\partial^2 u_2}{\partial x_2^2} \right| \Rightarrow x^* \text{ is locally stable}$$

$$\left| \frac{\partial^2 u_1}{\partial x_1 \partial x_2} \cdot \frac{\partial^2 u_2}{\partial x_1 \partial x_2} \right| > \left| \frac{\partial^2 u_1}{\partial x_1^2} \cdot \frac{\partial^2 u_2}{\partial x_2^2} \right| \Rightarrow x^* \text{ is not locally stable}$$

(where all these derivatives are taken at x).

Under the assumptions of the Corollary, the best reply sets BR_i are two C^1 curves that intersect at $x^\star$. The inequalities in (13) simply compare the modulus of the slopes $s_i = \dfrac{\partial^2 u_i}{\partial x_i \partial x_1} / \dfrac{\partial^2 u_i}{\partial x_i \partial x_2}$ to BR_i, $i = 1,2$:

$$|s_1| > |s_2| \Rightarrow x^\star \text{ is locally stable}$$

$$|s_1| < |s_2| \Rightarrow x^\star \text{ is not locally stable}$$

The IRS-game in Example 5 above gives an example of a non stable NE where $|s_1| < |s_2|$ whereas s_1 and s_2 have the same sign. If $|s_1| < |s_2|$ and s_1 , s_2 have opposite signs, we obtain the typical cob-web instability illustrated by Figure 7.

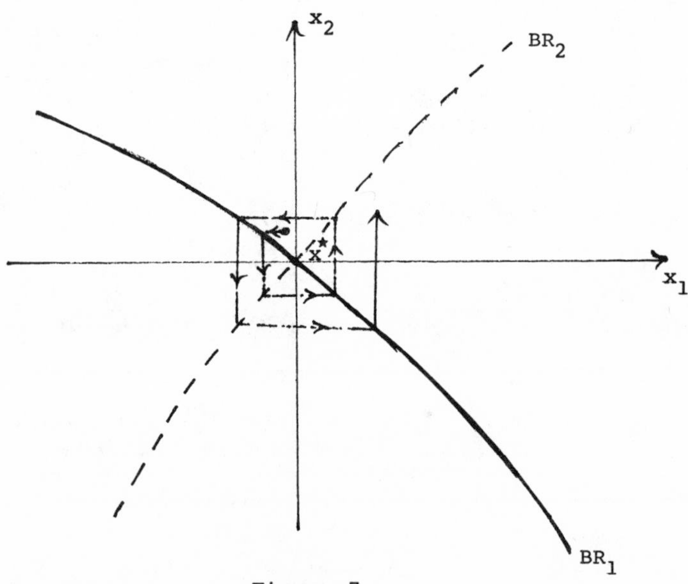

Figure 7

If on the contrary $|s_1| < |s_2|$, the locally stable N E $x^\star$ can be shown to be the unique sophisticated equilibrium of the restriction of the initial game to any sufficiently small rectangular neighboorhood of $x^\star$. Let us illustrate this point by one more Figure.

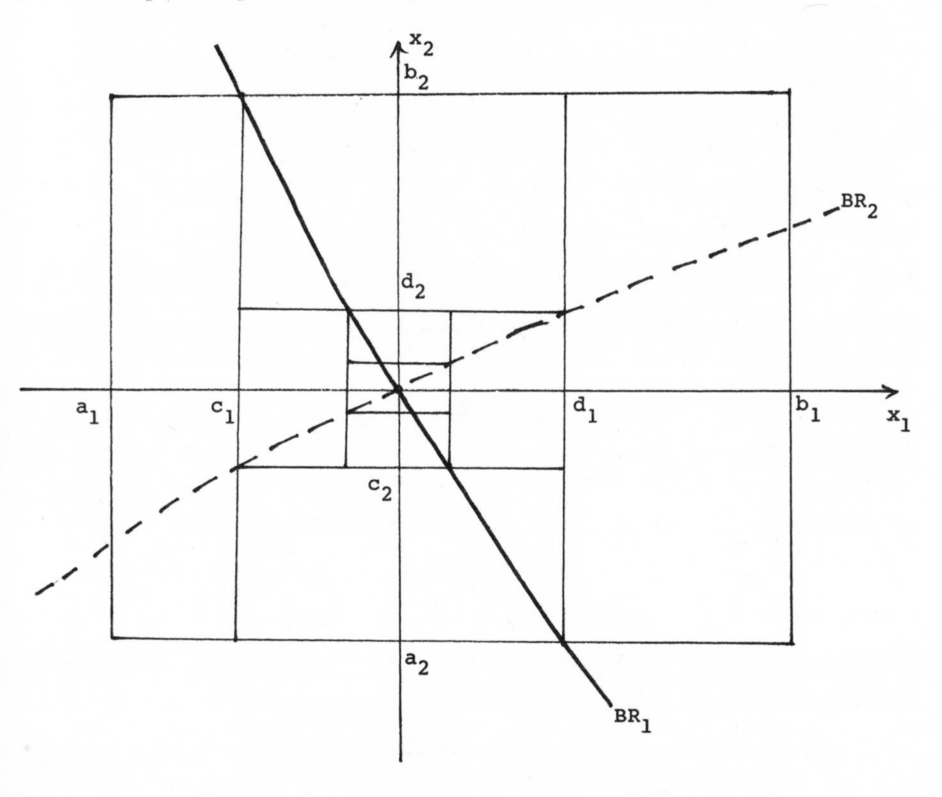

Figure 8

Starting with the game $([a_1,b_1],[a_2,b_2],u_1,u_2)$ we choose a_i, b_i so that u_i is concave w.r.t. x_i in $[a_1,b_1] \times [a_2,b_2]$. Then we observe that for all fixed $x_2 \in [a_2,b_2]$ the pay-off $u_2(\cdot, x_2)$ increases from a_1 until $r_1(x_2)$, next decreases until b_1. Since $r_1(x_2) \in [c_1,d_1]$ for all $x_2 \in [a_2,b_2]$ it follows that any strategy $x_1 \in [a_1,c_1]$ of player 1 is dominated by his strategy c_1, whereas any $x_1 \in [d_1,b_1]$ is dominated by d_1. Dropping player 1' dominated strategies leaves the game $([c_1,d_1],[a_2,b_2],u_1,u_2)$ in which, by a similar argument player 2' undominated strategies cover $[c_2,d_2]$ hence the reduced game $([c_1,d_1],[c_2,d_2],u_1,u_2)$. Performing this operation iteratedly clearly shrinks the strategy sets to the stable NE outcome x^*, as can be checked on Figure 8.

The connection between local stability and local dominance-solvability is actually very general. Say that the game G is <u>locally dominance-solvable at</u> x^* if there exists a rectangular neighboorhood V_N of x^* such that in the restriction of G to V_N, the successive elimination of dominated strategies shrinks V_N to x^* in the limit. Then under the assumption of Theorem 3, Gabay-Moulin [1980] prove that :

a) if the modulus of all eigenvalues of T is strictly less than 1, then G is locally dominance-solvable at x^*.

b) if G is locally dominance-solvable at $x^{\star}$ then the modulus of all eigenvalues of T is less than or equal to 1.

Remark : On Figure 8, we performed the elimination of dominated strategies "one player after the other" : by Lemma 4, Chapter II (to be adapted in the present framework) this is just the same as performing it simultaneously. As we have seen in Exercise 6 the same robustness does not hold for the Cournot tatonnement algorithm.

Exercise 7.

Pick a three player game where each strategy set is one dimensional. Give an almost necessary and sufficient condition for local $\{1,2,3\}$ - stability similar to Theorem 3.

Exercise 8.

In the deterministic case of the war of attrition (question 1 of Problem 5), compute the best response sets BR_i (which are not 1 dimensional curves) and analyze the stability of the Cournot tatonnement process. Pay special attention to the tatonnement starting at (0,0) and interpret it as an escalation scheme.

Next consider the variant of the war of attrition where a player cannot observe his opponent agressivity : strategy x_i by player i means that i is committed once and for all to remain agressive until time $t = x_i$. Henceforth the normal form game :

$$X_1 = X_2 = [0,+\infty[$$

$$u_1(x_1,x_2) = v_1 - x_1 \quad \text{if } x_2 < x_1$$

$$= -x_1 \quad \text{if } x_1 < x_2$$

$$= \frac{v_1}{2} - x_1 \quad \text{if } x_1 = x_2$$

with a similar definition for u_2.

Prove that this game has no NE outcome and analyze the Cournot tatonnement. Compare with the war of attrition.

Problem 5. Stability of a quantity-setting oligopoly.

We generalize Example 4 by taking now into account the production costs : $c_i(y)$ is the cost of producting y units of the good to producer i. Henceforth the game :

$$X_i = [0, +\infty[\quad u_i(x) = x_i . p(\bar{x}) - c_i(x_i), \text{ all } i \in N$$

On the function p we make assumptions (7). On the functions c_i we simply suppose :

$c_i' > 0$

Thus the production technologies can be either increasing or decreasing returns, or neither.

1) If each player's technology has decreasing or not too increasing returns in the following sense :

$$c_i''(y) > 2\,p'(y) + y\,p''(y) \quad \text{all } y \geqslant 0 \text{ , all } i = 1,\ldots,n. \tag{14}$$

show the existence of at least one N E outcome.

2) From now on we assume the existence of a N E outcome x^* without necessarily assuming (14). Prove that x^* is locally stable if the following inequalities hold :

$$c_i''(x_i^*) > (n-3)\,|\,p'(\bar{x}^*)\,| + (n-2)\,|\,p''(\bar{x}^*)\,|\,.\,x_i^*$$

Hint : Setting

$$p_i = \frac{p'(\bar{x}^*) + p''(\bar{x}^*)\cdot x_i^*}{2p'(\bar{x}^*) + p''(\bar{x}^*)\,.\,x_i^* - c_i''(x_i^*)}$$

prove that $0 < p_i < 1$ for all $i = 1,\ldots,n$ (taking into account $n \geqslant 2$). Next remark that an eigenvalue λ of T either is $-p_i$ for some $i = 1,\ldots,n$, or is a solution of :

$$\frac{p_1}{\lambda + p_1} + \ldots\ldots + \frac{p_n}{\lambda + p_n} = 1$$

3) If all players are identical ($c_i = c$ for all $i = 1,\ldots,n$), suppose that x^* is a N E on the diagonal :

$x_i^* = x_j^* = y^*$. Then prove the converse statement of the result in Question 2. If :

$$2p' + y^* . p'' < c'' < (n-3) \, |p'| + (n-2) . y^* \, |p''|$$

(where all p', p'' are taken at $\bar{x}^* = n\, y^*$ and c'' is taken at y^*), then x^* is not a stable NE outcome.

Notice that x^* being a NE outcome implies anyway : $2\,p' + y^* p'' \leqslant c''$ as u_i is maximal at x_i^* w.r.t. x_i.

Thus for $n = 2$ the stability of x^* is always true and becomes more and more difficult as n grows.

Problem 6.

1) In the auto dealer game (Problem 2) prove that the null NE outcome is not stable.

2) Turning to the non-null NE outcome, prove that the operator T (11) has always the eigenvalue 1. Using the same technique as in Problem 5 prove that this outcome is not locally stable if for at least two agents we have :

$$\frac{c_i}{p_i} < \frac{1}{2(n-1)} \left\{ \sum_{j=1}^{n} \frac{c_j}{p_j} \right\}$$

3) In the case $n = 2$ prove directly (i.e. by computing BR_1 and BR_2 and drawing a picture) that the non-null NE outcome is stable. Analyze similarly the stability

of the N E outcome in the quantity setting duopoly proposed in Exercize 4.

Problem 7. A continuous Cournot-tatonnement dynamic system. (*Moulin* [*1977*]).

Let $G = (X_i, u_i \; ; \; i \in N)$ be a n player game where each strategy set is one dimensional. Let φ be a realvalued function defined on $\mathbb{R}$ such that :

$$\varphi(0) = 0 \qquad \varphi'(t) > 0 \quad \text{all } t \in \mathbb{R}$$

Suppose that u_i is C^2 - differentiable on X_N for all $i \in N$ and consider the following differential equation on X_N :

$$\frac{dx_i}{dt} = \varphi(r_i(x_{\hat{i}}) - x_i) \quad \text{all } i \in N \tag{15}$$

1) Interpret this equation as a continuous version of a Cournot-tatonnement.

2) Any N E outcome x^* of G is a fixed point of (15). Find an almost necessary and sufficient condition (expressed by means of the operator T) for x^* to be a stable solution of (15) (i.e. for some neighboorhood V of x^*, system (15) starting in V never leaves V and eventually converges to x^*).

Prove in particular that if x^* is a locally stable N E outcome (Definition 3), it is a stable solution of (15). Give an example to show that the converse statement is not true.

Problem 8. Local prisonners' dilemma, *(Moulin [1979])*.

1) We fix an $n \times n$-matrix $A = [a_{ij}]_{i,j=1,\dots,n}$ and we assume $a_{ii} \neq 0$, all $i = 1,\dots,n$. Given $\lambda_1 > 0,\dots,\lambda_n > 0$ we denote by $G(A;\lambda)$ the game $(Y_i(\lambda_i), u_i, i = 1,\dots,n)$ where :

$$Y_i(\lambda_i) = [-\lambda_i, \lambda_i] \quad i = 1,\dots,n$$

$$\begin{bmatrix} u_1 \\ \cdot \\ \cdot \\ \cdot \\ u_n \end{bmatrix} = A \begin{bmatrix} x_1 \\ \cdot \\ \cdot \\ \cdot \\ x_n \end{bmatrix}$$

In $G(A;\lambda)$ each player has a unique dominating strategy and we denote by $x(\lambda) \in Y_N(\lambda)$ the dominating strategy equilibrium. Prove that the 4 following statements are equivalent :

i) For all $\lambda \gg 0$, $x(\lambda)$ is Pareto dominated in $G(A;\lambda)$

ii) For some $\lambda \gg 0$, $x(\lambda)$ is Pareto dominated in $G(A;\lambda)$

iii) There exists $x \in \mathbb{R}^n$ such that :

$$\left.\begin{array}{ll} u_i(x_i \cdot e_i) & > 0 \\ u_i(x) & < 0 \end{array}\right\} \quad \text{for all } i = 1,\ldots, n$$

where e_i is the i-th coordinate vector of $\mathbb{R}^n$.

iv) There exists no $p, q \geqq 0$ ($p_i, q_i \geqslant 0$, not all zero) such that

$$\sum_{i=1}^{n} p_i a_{ij} = q_j a_{jj} \quad , \quad \text{all } j = 1,\ldots, n$$

When these conditions hold, we say that the linear pay-offs A induce a Prisonners' dilemma.

2) We consider now a n-normal form game $G = (X_i , u_i ; i = 1,\ldots,n)$ where X_i are real open intervals and u_i are C^1-differentiable on X_N. Given an outcome $x \in X_N$ such that :

$$\frac{\partial u_i}{\partial x_i}(x) \neq 0 \qquad \text{all } i = 1,\ldots,n$$

we say that G has a local Prisonners' dilemma at x if $\dfrac{\partial(u_1,\ldots,u_n)}{\partial(x_1,\ldots,x_n)}$ induces a Prisonners' dilemma. Interpret this definition with the help of question 1. In the quantity setting oligopoly of Example 4 compute the outcomes at which a local Prisonners' dilemma arises.

3) In addition to the assumptions of question 2 we suppose $n = 2$. Prove that G has a local prisonners'

dilemma at x if and only if the following system holds :

$$\frac{\partial u_i}{\partial x_i} \cdot \frac{\partial u_j}{\partial x_i} < 0 \qquad \text{all } \{i, j\} = \{1, 2\}$$

$$\frac{\partial u_1}{\partial x_1} \cdot \frac{\partial u_2}{\partial x_2} \cdot \left(\frac{\partial u_1}{\partial x_1} \cdot \frac{\partial u_2}{\partial x_2} - \frac{\partial u_1}{\partial x_2} \cdot \frac{\partial u_2}{\partial x_1}\right) < 0 \qquad (16)$$

all the derivatives being taken at x.

Prove that in general any neighboorhood of a N E outcome or of a Pareto optimum intersects area (16) (give a precise meaning to "in general").

By drawing a figure of area (16) on the two oligopoly models of Example 5, show that area (16) lies "in between" the N E outcome and the Pareto optimum line. To what extent can you say that this configuration is general ? How are the condition (16) generalized to the case $n = 3$? to any integer value of n ?

REFERENCES

BERGE, C. 1957. Espaces véctoriels topologiques. Paris : Dunod.

CASE, J.H. 1979. Economics and the competitive process. New York : New York University Press.

GABAY, D. and H. MOULIN. 1980. "On the uniqueness and stability of Nash equilibrium in non cooperative games" in Applied stochastic control in econometric and management science, Bensoussan, Kleindorfer, Tapiero Eds., Amsterdam : North-Holland Publishing Co.

GROTE, J. 1974. "A global theory of games". Journal of Mathematical Economics 1, 3 : 223-236.

ORTEGA, J.M. and W.C. RHEINBOLDT. 1970. Iterative solution of non-linear equations in several variables. New York : Academic Press.

MILGROM, P. and R.J. WEBER. 1980. "Distributional strategies for games with incomplete information". The Center for Mathematical Studies in Economics and Management Science, Northwestern University.

MOULIN, H. 1977. "On the asymptotic stability of agreements", in Systèmes dynamiques et modèles économiques, G. Fuchs ed., Paris : C.N.R.S..

MOULIN, H. 1979. "Two and three person games : a local study" in International Journal of Game Theory 8, 2 : 81-107.

MOULIN, H. 1981. "Deterrence and cooperation". European Economic Review 15 : 179-193.

NASH, J.F. 1951. "Non cooperative games". Annals of Maths.54 : 286-295.

OKUGUCHI, K. 1976. Expectations and stability in oligopoly models. Lect.Notes in Eccs.and Math.Systems 138. Springer Verlag.

RAND, D. 1978. "Exotic phenomena in games and duopoly models". Journal of Mathematical Economics 5, 2 : 173-184.

ROSEN, J.B. 1965. "Existence and uniqueness of equilibrium points for concave N-person games". Econometrica 33 : 520-533.

SCHELLING, T.C. 1979. Micromotives and macro-behaviour. New York : Norton Publ.

SCHELLING, T.C. 1971. The strategy of conflict. Cambridge (USA): Harvard University Press.

CHAPTER IV. MIXED STRATEGIES.

In some normal form games although each player can pick at will any element from his or her strategy set, a rational tactical move is to incorporate a voluntary randomness within this choice. This creates some uncertainty about my own strategy and the induced reaction by the others can prove eventually profitable (to me).

Randomization of one's behaviour requires privacy of the actually chosen strategy (i.e. the players must pick a strategy independently and simultaneously). It modelises bluffing and captures the idea of ubiquity in optimal strategies : "A ruse is to pretend that I am doing this here whereas I am doing that there. Thus it amounts to potential ubiquity, as the early strategical authors noticed already. E.g. Xenophon : to know that there is a post somewhere while ignoring its position and strength, removes all security and makes every place suspect". (Guilbaud [1968])

Allowing mixed (i.e. random) strategies together with the assumption that the players are endowed with von Neumann - Morgenstern utility functions, leads to extend the original game into its mixed version. In doing so, new Nash equilibrium outcomes emerge.

When the initial strategy sets are finite the mixed extension of a game always has a non-empty N E set (Section 1), and the mixed NE set can be computed in many cases (Section 2).

When the initial strategy sets are infinite, specific topological difficulties arise. Compactness of the strategy sets and continuity of the pay-off functions are nevertheless sufficient to guarantee the existence of at least one N E outcome (Section 3).

I. THE MIXED EXTENSION OF A GAME

Example 1. de Montmort's game.

In the late 18th century, the french mathematician René de Montmort considers the following situation : to make a gift to his son, a father proposes : "I shall have a gold coin into my right hand or my left hand and you shall name a hand ; if your guess is correct and it is my right hand you shall get the gold coin ; if your guess is correct and it is my left hand, you shall receive _two_ gold coins ; otherwise you shall get nothing." Then Montmort ask how much this gift is actually worth to the son while pointing out that "in this game if the players are equally shrewd and clear-sighted ; there is no way we can lay down a rule of conduct", i.e. no optimal strategy exists in this game*.

---

* In fact, a contemporary of Montmort, James Waldegrave, did proposed to use mixed strategies in a similar game. It is not clear, however, that he had the intuition of optimal mixed strategies (see the historical article by Rives [1975]).

Our 2 × 2 two-person zero-sum game is then :

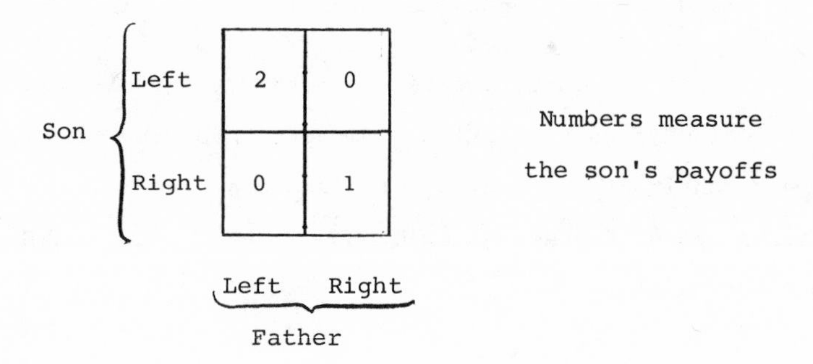

As this game has no value ($\sup_{x_1} \inf_{x_2} u = 0$, $\inf_{x_2} \sup_{x_1} u = 1$), no player has an optimal strategy. As in any 2-person zero-sum game without value, discovering which strategy the opponent will use allows one to win by optimally replying (see Chapter I, Section 3). Hence a struggle for the <u>second</u> move, where each player wants to conceal his final strategic choice, and at the same time spies the other's intention. But the deepest secrecy is not enough to prevent my opponent from guessing my strategic choice, if I am to settle this choice on deterministic deductive arguments, for my rational opponent could always reconstruct this very line of argument (see Remark 1, Section 1, Chapter III).

The natural way to prevent one's own choice from being guessed is to make this choice at random : instead of choosing a strategy, so called pure, x_1 within {Left , Right}, the son will use a randomized strategy μ_1 taking the value Left or

Right with respective probability p_1 , $1 - p_1$. Imagine for instance that he designs a coin that shows L with probability p_1 and tosses this coin to choose his actual strategy. Assume now that the father can never observe the actual draw of the coin. This implies that the father's strategy, whether or not it is a random variable, can not be correlated to the son's strategy and therefore the expected payoff to the son of the strategy, so called mixed, μ_1 is at least :

$$\inf \{2 p_1 , 1 - p_1\}$$

By choosing $p_1 = \frac{1}{3}$, the son is guaranteed of an expected payoff <u>at least</u> $\frac{2}{3}$. Observe now that the situation of the father is symmetrical : by playing at random strategies Left and Right with respective probabilities p_2 , $1 - p_2$, his maximal expected loss is :

$$\sup \{2 p_2 , 1 - p_2\}$$

Choosing $p_2 = \frac{1}{3}$, the father is guaranteed of an expected loss <u>at most</u> $\frac{2}{3}$. We conclude that the possibility of incorporing tactical uncertainty into their strategical choices suggests the value $\frac{1}{3}$ as a fair estimate of the father's generosity.

Exercise 1.

Use a similar argument for the tic-tac-toe game (Example 4, Chapter I). Among players using randomized strategies, prove that the value of the game is $\frac{6}{11}$.

Definition 1.

Let $G = (X_i , u_i ; i \in N)$ be a N-normal form game where X_i is a finite set for all $i \in N$.

A mixed strategy of player i is a probability distribution μ_i over X_i. Hence the set M_i of player i's mixed strategies is the unit simplex of $\mathbb{R}^{X_i}$

The mixed extension of G is the N-normal form game $G_m = (M_i , \bar{u}_i ; i \in N)$ where

$$\text{for all } \mu \in M_N : \bar{u}_i(\mu) = \sum_{x \in X_N} u_i(x)\ \mu_1(x_1) \cdot \mu_2(x_2) \cdot \dots \cdot \mu_n(x_n) \qquad (1)$$

In the mixed extension G_m of G, player i's strategy is a probability distribution μ_i over X_i. It is understood that player i actually constructs a private lottery that selects a strategy $x_i \in X$ according to μ_i. Privacy of the lottery means that player i alone is informed of the strategy x_i actually selected by the lottery (even if the other players may know the probability distribution μ_i). Moreover i's lottery is stochastically independent from j's lottery, for all j, $j \neq i$ (so that j can deduce no information on i's lottery by observing his own lottery).

As the random variables are mutually independent, $\bar{u}_i(\mu)$ is the expected utility of player i. Taking $\bar{u}_i$ as the payoff function of i in G_m amounts to say that player i compares various lotteries μ , μ' on X_N by simply comparing their associated expected utility $\bar{u}_i(\mu)$, $\bar{u}_i(\mu')$. In other words u_i is a von Neumann - Morgenstern utility function that summarizes the preferences of agent i over all conceivable lotteries on X_N. More justifications of this assumption pertains to the statistical theory of decision and will not be pursued here : see Luce and Raiffa [1957]. Notice that the cardinality of $u_i(x)$ plays a fundamental role here whereas in the three previous chapters, only the ordinal preference orderings induced by the u_i's over X_N did matter.

A pure strategy $x_i \in X_i$ of player i in the initial game G will be identified with the mixed strategy $\delta_{x_i} \in M_i$ which selects x_i with probability 1 :

$$\begin{cases} \delta_{x_i}(x_i') = 0 \quad \text{all } x_i' \in X_i \text{ , } x_i' \neq x_i \\ \delta_{x_i}(x_i) = 1 \end{cases}$$

Namely, formula (1) implies at once :

$$\bar{u}_i(\delta_x) = u_i(x) \quad \text{for all } x \in X_N \text{ and } \delta_x = (\delta_{x_i})_{i \in N}$$

Thus we look at X_i as a subset of M_i and at $\bar{u}_i$ as the extension of u_i from X_N to M_N.

Theorem 1.

If X_i is finite for all $i \in N$, the set of Nash equilibrium outcomes of G_m is a non-empty compact subset of M_N. Moreover it contains the set of Nash equilibrium outcomes of G :

$$NE(G) \subset NE(G_m) \neq \emptyset \qquad (2)$$

Proof.

Pick any NE outcome x of G. Denoting

$$\delta_{x_{\hat{i}}} = (\delta_{x_j})_{j \neq i} \in M_{N \setminus \{i\}}$$

we observe that :

$$\sup_{\mu_i \in M_i} \bar{u}_i \ (\mu_i \ , \delta_{x_{\hat{i}}}) = \sup_{y_i \in X_i} u_i \ (y_i \ , x_{\hat{i}})$$

Since M_i is the convex hull of X_i (when the latter is identified with a subset of M_i) and $\bar{u}_i$ is linear with respect to its variable μ_i. Now by the NE property of x :

$$\sup_{y_i \in X_i} u_i \ (y_i \ , x_{\hat{i}}) = u_i(x) = \bar{u}_i \ (\delta_{x_i} \ , \delta_{x_{\hat{i}}})$$

Combining these properties yields that δ_x is a NE outcome of G_m.

Remark next that G_m satisfies all the assumptions of Nash's theorem (Theorem 2, Chapter III). Namely M_i is convex and compact within $\mathbb{R}^{X_i}$, $\bar{u}_i$ is continuous over M_N and linear with respect to its variable μ_i. This concludes the proof of Theorem 1.

■

As a consequence of Lemma 1, Chapter III, all NE outcomes of the mixed game G_m are individually rational in G_m. Actually they are also individually rational in the original game G as our next result implies :

<u>Lemma 1</u>.

The secure utility level of player i in the original game is less than or equal to his secure utility level in the mixed game :

$$\text{for all } i \in N : \sup_{x_i \in X_i} \inf_{x_{\hat{i}} \in X_{\hat{i}}} u_i(x_i, x_{\hat{i}}) \leq \sup_{\mu_i \in M_i} \inf_{\mu_{\hat{i}} \in M_{\hat{i}}} \bar{u}(\mu_i, \mu_{\hat{i}}) \qquad (3)$$

Proof.

Fix a player $i \in N$, a pure strategy $x_i \in X_i$ and some mixed strategies $\mu_j \in M_j$ for all $j \in N \setminus \{i\}$. Then $\bar{u}_i(\delta_{x_i}, \mu_{\hat{j}})$ is the expected value of the function $u_i(x_i, .) \in \mathbb{R}^{X_{\hat{i}}}$ (the restriction of u_i to $X_{\hat{i}}$ for fixed x_i) with respect to the product probability measure $\bigotimes_{j \in N \setminus \{i\}} \mu_j$ on $X_{\hat{i}}$.

Therefore we have :

$$\inf_{x_{\hat{i}} \in X_{\hat{i}}} u_i(x_i, x_{\hat{i}}) \leqslant \bar{u}_i(\delta_{x_i}, \mu_{\hat{i}})$$

The above inequality being true for all $\mu_{\hat{j}} \in M_{\hat{j}}$ we get :

$$\inf_{x_{\hat{i}} \in X_{\hat{i}}} u_i(x_i, x_{\hat{i}}) \leqslant \inf_{\mu_{\hat{i}} \in M_{\hat{i}}} \bar{u}_i(\delta_{x_i}, \mu_{\hat{i}}) \leqslant \sup_{\mu_i \in M_i} \inf_{\mu_{\hat{i}} \in M_{\hat{i}}} \bar{u}_i(\mu_i, \mu_{\hat{i}})$$

Since the choice of x_i was arbitrary the proof of Lemma 1 is complete.

■

When applied to zero-sum games the above results yield :

Corollary of Theorem 1 and Lemma 1.

Given a two-person zero-sum game $G = (X_1, X_2, u_1)$ where X_1, X_2 are both finite, its mixed extension is a two person-zero-sum game $G_m = (M_1, M_2, \bar{u}_1)$. The game

G_m has (at least) a saddle-pair and a value, that we call the <u>mixed value</u> of G and denote by $v_m(u_1)$.

Moreover :

$$\sup_{x_1 \in X_1} \inf_{x_2 \in X_2} u_1 (x_1, x_2) \leqslant v_m(u_1) \leqslant \inf_{x_2 \in X_2} \sup_{x_1 \in X_1} u_1 (x_1, x_2) \tag{4}$$

Suppose first that the initial game G has a value and each player has at least one optimal strategy in G. Then the mixed extension of G has the same value and the optimal strategies of both players in G_m are essentially any convex combination of optimal strategies in G (see Problem 2, question 1).

Suppose on the contrary that G has no value, and therefore no player has an optimal strategy. Then G_m endows each player with at least one optimal mixed strategy and its value lies within the "duality gap" $[\sup_{x_1} \inf_{x_2} u_1 , \inf_{x_2} \sup_{x_1} u_1]$. (With in general a strict inequality on both sides of (4), see Problem 1 below). A typical example is Montmort's game (Example 1) with mixed value 1/3 and optimal prudent strategies ($\frac{1}{3}$ Left + $\frac{2}{3}$ Right) for both players. See also Section 2, in particular Exercizes 7, 8.

To illustrate further Theorem 1, consider the prisonners' dilemma (Example 1, Chapter I). There each player's agressive strategy is still his unique dominating strategy in the mixed extension of the game, so that :

$$NE(G) = NE(G_m)$$

Our last example illustrates a situation where :

$$\emptyset \neq NE(G) \underset{\neq}{\subset} NE(G_m)$$

Example 2. The mixed crossing game.

The crossing game (Example 2, Chapter III) has two NE outcomes in pure strategies. Its mixed version has one more NE outcome, namely the pair $(\mu_1^\star , \mu_2^\star)$

$$\mu_1^\star = \mu_2^\star = \frac{1 - \varepsilon}{2 - \varepsilon} \text{ Stop} + \frac{1}{2 - \varepsilon} \text{ Go} \quad (5)$$

Namely we have :

$$\bar{u}_1 \; (\delta_{Stop} , \mu_2^\star) = \frac{1 - \varepsilon}{2 - \varepsilon} \times 1 + \frac{1}{2 - \varepsilon} \times (1 - \varepsilon) = 1 - \frac{\varepsilon}{2 - \varepsilon}$$

$$\bar{u}_1 \; (\delta_{Go} , \mu_2^\star) = \frac{1 - \varepsilon}{2 - \varepsilon} \times 2 + \frac{1}{2 - \varepsilon} \times 0 = 1 - \frac{\varepsilon}{2 - \varepsilon}$$

Since $\bar{u}_1$ is linear with respect to μ_1 we deduce :

$$\bar{u}_1 \; (\mu_1 , \mu_2^\star) = \bar{u}_1 \; (\mu_1^\star , \mu_2^\star) \quad \text{for all } \mu_1 \in M_1 .$$

Exchanging the roles of players 1 and 2 leads to a symmetrical computation :

$$\bar{u}_2 (\mu_1^\star , \mu_2) = \bar{u}_2 (\mu_1^\star , \mu_2^\star) \qquad \text{for all } \mu_2 \in M_2.$$

The two above properties imply that $(\mu_1^\star , \mu_2^\star)$ is a NE outcome of the mixed crossing game. Actually they say more : given that player j uses $\mu_j^\star$, all player i's strategies are indifferent to him. Thus the equilibrium strategy of player i supplies the exact level of uncertainty that makes all moves of player j indifferent (to j). This feature is not uncommon : in Section 2 below, we show that it is a necessary property of all completely mixed NE outcomes, i.e. those outcomes where the probability that a given player uses a given strategy is never zero. Not only are completely mixed equilibria easily interpreted ; they also are simply computed (see Lemma 3, Section 2).

For the time being we observe that, contrary to the two NE outcomes in pure strategies, the (completely) mixed NE outcome is Pareto dominated (its payoff vector $(1 - \frac{\varepsilon}{2 - \varepsilon} , 1 - \frac{\varepsilon}{2 - \varepsilon})$ is dominated by (1 , 1)). However $(\mu_1^\star , \mu_2^\star)$ is symmetrical $(\mu_1^\star = \mu_2^\star)$ and equitable $(u_1^\star = u_2^\star)$ and could therefore be recommended on normative grounds.

Existence of a symmetrical mixed NE in symmetrical games actually is a general fact :

Exercise 2.

Let $(X_i \ , u_i \ ; \ i \in N)$ be a symmetrical game :

$$\begin{cases} X_i = X_j \quad \text{all } i, j \in N \\ u_i \ (x_i, x_{N\setminus\{i,j\}}, x_j) = u_j \ (x_j, x_{N\setminus\{i,j\}}, x_i) \quad \text{all } i, j \in N \\ \qquad\qquad\qquad\qquad\qquad\qquad\qquad \text{all } x \in X_N \end{cases}$$

Prove that its mixed extension has at least one symmetrical NE outcome μ :

$$\mu_i = \mu_j \quad \text{all } i \ , \ j \in N$$

Hint :

Follow the proof of Theorem 2, Chapter II by setting $\phi \ (\mu \ , \ \nu) = u_1 \ (\mu_1 \ , \ \nu_{\hat{1}}) - u_1(\nu)$ all $\mu \ , \ \nu \in M_N$.

Exercise 3. Where the NE strategy is not prudent.

We already know (Examples 2 and 3, Chapter III) that a NE strategy might be a non prudent one. Here we show that this can also happen in mixed games. Consider the 2×2 game G :

player 1	0 / 1	2 / 0
	2 / 0	1 / 1

player 2

Prove that in the mixed extension of G, each player has a unique prudent strategy μ_i^P and there is a unique N E outcome (μ_1^N, μ_2^N).

Check the following relations :

$$u_1 (\mu_1^N, \mu_2^P) < u_1 (\mu_1^N, \mu_2^N) = u_1 (\mu_1^P, \mu_2^N) = u_1 (\mu_1^P, \mu_2^P)$$

Interpretation ?

Exercise 4. An example of bluffing.

With a uniform probability, player 1 secretely draws a card marked High or Low. Then he decides either to give a dollar to player 2 or to call ; in the latter case player 2 decides either to swear, in which case he gives a dollar to player 1, or to call, in which case player 1's card is shown up and 2 pays 4 dollars (to 1) if the card is High whereas 1 pays 5 dollars (to 2) if the card is Low.

Model this two-person zero-sum game in normal form and compute the optimal mixed strategies of both players and the mixed value.

<u>Exercise 5</u>. <u>Two-person zero-sum game and linear programming.</u>

Given $G = (X_1 , X_2 , u_1)$ prove that the linear programming problem :

$$\max \alpha$$

$$\text{s.t. } \mu_1 \in M_1 \text{ and } \alpha \leqslant \bar{u}_1 (\mu_1 , \delta_{x_2}) \text{ all } x_2 \in X_2 \tag{6}$$

(where the unknown variable is (α , μ_1)), gives the mixed value of G and player 1's optimal mixed strategies. What is the dual problem of (6) ?

Problem 1. *Mixed strategies in two person zero-sum games are profitable to both players.*

Let (X_1 , X_2 , u_1) be a two person zero-sum game with no value :

$$\sup_{x_1} \inf_{x_2} u_1 = a_1 < a_2 = \inf_{x_2} \sup_{x_1} u_1$$

1) Suppose that for all fixed $x_i \in X_i$ the mapping $x_j \to u_1$ is one-to-one over X_j (all $\{i , j\} = \{1 , 2\}$). Prove that the mixed value $v_m(u_1)$ belongs to the <u>open</u> interval $]a_1 , a_2[$.

2) Give an example (violating the one-to-one assumption) where

$$\sup_{x_1} \inf_{x_2} u_1 = v_m(u_1) < \inf_{x_2} \sup_{x_1} u_1$$

Problem 2.

1) If the two-person zero-sum game $G = (X_1, X_2, u_1)$ has a value, prove that any convex combination of optimal strategies in G is an optimal strategy in G_m. Give an example to show that the converse is not true.

2) Let now $G = (X_i, u_i ; i \quad N)$ be a N-normal form game. Prove by an example that an undominated pure strategy x_i of player i in G ($x_i \in \mathcal{D}_i(u_i)$) might be a dominated strategy in the mixed game G_m.

3) Prove that if x_i is a dominated strategy of player i in G ($x_i \notin \mathcal{D}_i(u_i)$) then the mixed game G_m has at least one NE outcome μ where player i does not use strategy x_i :

$$\mu_i(x_i) = 0$$

4) Prove that if player i has a dominating strategy μ_i in G_m ($\mu_i \in D_i(\bar{u}_i)$) then, all pure strategies x_i such that $\mu_i(x_i) > 0$ are equivalent, and actually are dominating in G ($x_i \in D_i(\bar{u}_i)$).

Problem 3. *Exchangeability of mixed NE outcomes in two-person games.* (Parthasarathy-Raghavan [1971]),

In a two-person mixed game $G_m = (M_1, M_2, u_1, u_2)$ the NE outcomes are not in general exchangeable :

$$\mu, \mu' \in NE(G_m) \not\Rightarrow (\mu_1, \mu_2'), (\mu_1', \mu_2) \in NE(G_m)$$

This should be clear from the mixed crossing game (Example 2 above).

1) Prove that the NE outcomes of G_m are exchangeable if and only if $NE(G_m)$ is a convex subset of $\mathbb{R}^{X_1} \times \mathbb{R}^{X_2}$ (where X_i are, of course, finite).

2) If $NE(G_m)$ is indeed a convex (hence rectangular) subset of $M_1 \times M_2$, prove the existence of at least one mixed NE outcome $\mu^\star$ that Pareto dominates, or has the same payoff vector as, all other mixed NE outcomes :

$$\forall \mu \in NE(G_m) \quad u_i(\mu) \leq u_i(\mu^\star) \quad i = 1, 2.$$

II. COMPUTATION OF THE MIXED NASH EQUILIBRIA.

Definition 2.

Let X_i be the finite strategy set of player i, $i \in N$. For any mixed strategy $\mu_i \in M_i$, we denote by $[\mu_i]$ the carrier of μ_i, i.e. the set of player i's pure strategies with a strictly positive probability in μ_i :

$$[\mu_i] = \{x_i \in X_i \ / \ \mu_i(x_i) > 0\}$$

We say that a mixed strategy μ_i is completely mixed if its carrier is the whole (pure) strategy set : $[\mu_i] = X_i$. We say that an outcome μ of G_m is completely mixed if μ_i is completely mixed for all $i \in N$.

Theorem 2.

Let the initial game $G = (X_i , u_i ; i \in N)$ be given with finite strategy sets and let $\mu \in NE(G_m)$ be a mixed Nash equilibrium. Then the following system holds true :

$$\forall i \in N \quad \forall x_i \in [\mu_i] \quad u_i (\delta_{x_i} , \mu_{\hat{i}}) = u_i(\mu) \qquad (7)$$

Proof.

Pick an agent $i \in N$. By the NE property we have :

$$u_i (\delta_{x_i} , \mu_{\hat{i}}) \leqslant u_i(\mu) \qquad \text{all } x_i \in [\mu_i] \qquad (8)$$

Suppose at least one of these inequalities is strict :

$$u_i(\delta_{x_i^o}, \mu_{\hat{i}}) < u_i(\mu)$$

Multiplying each inequality in (8) by $\mu_i(x_i)$ and taking into account that $\mu_i(x_i^o) > 0$ we get :

$$u_i(\mu) = u_i\left(\sum_{x_i \in [\mu_i]} \mu_i(x_i) \cdot \delta_{x_i}, \mu_{\hat{i}}\right) =$$

$$= \sum_{x_i \in [\mu_i]} \mu_i(x_i) \cdot u_i(\delta_{x_i}, \mu_{\hat{i}}) < \left\{\sum_{x_i \in [\mu_i]} \mu_i(x_i)\right\} \cdot u_i(\mu) = u_i(\mu)$$

a contradiction. Hence all inequalities in (8) actually are equalities.

■

Theorem 2 states that at a mixed Nash equilibrium μ player i's best replies to $\mu_{\hat{i}}$ contains <u>all</u> mixed strategies μ_i' with the same carrier as μ_i. In particular, if μ is a completely mixed equilibrium, then <u>any</u> strategy of player i is a best reply to the $N \setminus \{i\}$-uple $\mu_{\hat{i}}$ of equilibrium strategies by the other players. Thus at a completely mixed equilibrium of a <u>two</u>-person game, a player i picks a strategy that makes player j indifferent among all his (mixed) strategies, quite independently of his own payoff u_i. This is illustrated by Example 2 above and Lemma 3 below. However, in a game with at least <u>three</u> players the completely mixed equilibrium strategies must be determined jointly, so that μ_i depends on the payoff function u_i via system (7).

Theorem 2 is the key ingredient for the computation of mixed Nash equilibrium outcomes. Namely suppose that we seek for a mixed NE outcome $\mu \in NE(G_m)$ with a given carrier for all μ_i :

$$[\mu_i] = Y_i \subset X_i \qquad \text{all } i \in N \tag{9}$$

Then system (7) is equivalently rewritten as :

$$\forall\, i \in N \quad \forall\, x_i, y_i \in Y_i \quad \bar{u}_i\,(\delta_{x_i}, \mu_{\hat{i}}) = \bar{u}_i\,(\delta_{y_i}, \mu_{\hat{i}}) \tag{10}$$

This provides $\sum_{i \in N} (|Y_i| - 1)$ independent 1-dimensional equations. By (9) each μ_i consists of $(|Y_i| - 1)$ independent unknown variables (taking the constraint $\sum_{x_i \in Y_i} \mu_i(x_i) = 1$ into account) and we are home. A rigorous genericity argument can be developed (see e.g. Moulin [1981], Chapter IV) that leads to the following statement.

Fix X_i , $i \in N$; then there exist an open dense subset Ω of $\left(\mathbb{R}^{X_N}\right)^N$ such that for all $(u_i)_{i \in N}$ within Ω and all rectangular subsets $Y_N = \times_{i \in N} Y_i$ of X_N, the system (9) (10) with unknown $\mu_i \in M_i$, $i \in N$ has <u>at most</u> one solution.

Since X_N has finitely many rectangular subsets we conclude that when $(u_i)_{i\in N}$ belongs to Ω, the corresponding mixed game G_m has <u>finitely many</u> NE outcomes :

$$NE(G_m) \text{ is in general finite}$$
$$(\text{i.e. when } (u_i)_{i\in N} \text{ belongs to } \Omega)$$

When the strategy sets X_i , $i \in N$ have (very) few elements, the following algorithm allows exhaustive computation of $NE(G_m)$: fix a rectangular subset Y_N of X_N. Solve (9) (10) with $\mu_i \in M_i$, all $i \in N$.

If μ is such a solution, check the remaining inequalities :

$$\forall\, i \in N \quad \forall\, x_i \in X_i \setminus Y_i : \bar{u}_i\,(\delta_{x_i}\,,\,\mu_{\hat{i}}) \leq \bar{u}_i\,(\mu_i\,,\,\mu_{\hat{i}})\,.$$

<u>The case of bimatrix games</u>.

For two-person games the above computations are made more transparent by the matrix representation that we already use to describe simple examples. Player i's payoff is now a matrix

$$U_i = [u_i\,(x_1\,,\,x_2)]_{\substack{x_1\in X_1\\ x_2\in X_2}}$$

where X_1 is the set of rows and X_2 the set of columns.

A mixed strategy $\mu_1 \in M_1$ is a <u>row</u> vector

$$\mu_1 = [\mu_1(x_1)]_{x_1 \in X_1}$$

whereas a mixed strategy $\mu_2 \in M_2$ is a <u>column</u> vector

$$\mu_2 = [\mu_2(x_2)]_{x_2 \in X_2}$$

Then the mixed payoff of player i is the ordinary matrix product :

$$\bar{u}_i (\mu_1 , \mu_2) = \mu_1 U_i \mu_2$$

Suppose now that (μ_1 , μ_2) is a <u>completely mixed</u> NE outcome. Then system (10) becomes :

$$\begin{aligned} U_1 \mu_2 &= v_1 \mathbb{1}_{X_1} \\ \mu_1 U_2 &= v_2 \mathbb{1}_{X_2} \end{aligned} \qquad (11)$$

where $v_i = \mu_1 U_i \mu_2$ is player i's NE payoff and $\mathbb{1}_{X_1}$ (resp $\mathbb{1}_{X_2}$) is the X_1-column vector (resp. the X_2-row vector) with all components equal to one.

The second equation of system (11) says that the $|X_2|$ column vectors or U_2 all belong to the same affine subspace of $\mathbb{R}^{X_1}$, namely :

$$A = \{z \in \mathbb{R}^{X_1} / \mu_1 . z = v_2\}$$

If $|X_1| < |X_2|$ this yields a genuine dependency of these column vectors (i.e. one can eliminate μ_1 , v_2 from the second equation of (11)). Therefore in general (i.e. on an open dense subset of $X_1 \times X_2$ matrices) $|X_1|$ must be at least $|X_2|$ and by a symmetrical argument we conclude $|X_1| = |X_2|$.

Supposing now that (μ_1 , μ_2) is any mixed NE outcome we can develop the same argument for the restriction of the matrices U_i , $i = 1 , 2$, to $[\mu_1] \times [\mu_2]$. Hence the following result.

<u>Lemma 2</u>.

For any given (finite) X_1 , X_2 , there is an open dense subset Ω_o of $\mathbb{R}^{X_1 \times X_2}$, the set of $X_1 \times X_2$ - matrices, such that for all U_1 , U_2 in Ω_o we have :

$$(\mu_1 , \mu_2) \in NE (U_1 , U_2) \Rightarrow |[\mu_1]| = |[\mu_2]|$$

As another application of system (11) we can actually compute the equilibrium strategies and equilibrium payoffs as soon as the corresponding carriers are known. Typically :

<u>Lemma 3</u>.

Suppose $|X_1| = |X_2|$ and that U_1 , U_2 are two regular $X_1 \times X_2$ matrices. Then if the game $G = (X_1 , X_2 , U_1 , U_2)$ has a completely mixed NE outcome (μ_1 , μ_2) it is uniquely given by :

$$\left.\begin{array}{l}\mu_1 = v_2 \cdot \mathbb{1}_{X_2} U_2^{-1} \\ \\ \mu_2 = v_1 U_1^{-1} \mathbb{1}_{X_1}\end{array}\right\} \text{ and } v_i = \frac{1}{\mathbb{1}_{X_2} U_i^{-1} \mathbb{1}_{X_1}} \qquad i = 1, 2 \qquad (12)$$

Conversely, if the vectors μ_1 , μ_2 given by (12) have all their components non negative, the pair (μ_1 , μ_2) is a mixed NE outcome of G.

<u>Proof</u>.

Multiplying the first equation in (11) by U_1^{-1} yields : $\mu_2 = v_1 U_1^{-1} \mathbb{1}_{X_1}$. Multiplying the latter by $\mathbb{1}_{X_2}$ gives $1 = \mathbb{1}_{X_2} \cdot \mu_2 = v_1 \mathbb{1}_{X_2} U_1^{-1} \mathbb{1}_{X_1}$, hence v_1 is non zero and equals $\frac{1}{\mathbb{1}_{X_2} U_1^{-1} \mathbb{1}_{X_1}}$. The converse statement is proved just as easily: from (12) we deduce (11) and therefore

$$\begin{cases}\mu_1 U_1 \mu_2 = \mu_1' U_1 \mu_2 & \text{all } \mu_1' \in M_1 \\ \mu_1 U_2 \mu_2 = \mu_1 U_2 \mu_2' & \text{all } \mu_2' \in M_2\end{cases}$$

Notice that (μ_1 , μ_2) is a completely mixed NE outcome only if all its components are <u>strictly</u> positive, which may not be the case from system (12).

∎

For the sake of computation, it is easier to read the equilibrium payoffs in (12) as :

$$v_i = \frac{\det(U_i)}{\sum_{\substack{x_1 \in X_1 \\ x_2 \in X_2}} U_i(x_1, x_2)} \tag{13}$$

where $U(x_1, x_2)$ stands for the cofactor of entry (x_1, x_2) in matrix U.

Example 3. 2 × 2 bimatrix games.
(each player has two strategies).

We consider the following family of bimatrix games

player 1		Left	Right
	Top	a_1 , a_2	c_1 , c_2
	Bottom	b_1 , b_2	d_1 , d_2

player 2

We suppose that a_i , b_i , c_i , d_i are four distinct real numbers ($i = 1, 2$). This is enough to avoid degenerate matrices: the set $NE(G_m)$ is always finite. Exactly three different cases arise :

<u>Case 1</u> :

<u>In the initial game G, at least one player</u>, say player 1, <u>has a dominant strategy, say Top</u>.

Then G and its mixed extension G_m have a unique Nash equilibrium :

$$NE(G) = NE(G_m) = \{(Top\ ,\ Left)\} \quad \text{if } a_2 > c_2$$

$$= \{(Top\ ,\ Right)\} \quad \text{if } a_2 < c_2$$

Namely inequalities $a_1 > b_1$, $c_1 > d_1$ imply that in G_m , the pure strategy Top strictly dominates all other mixed strategies.

<u>Case 2</u> :

<u>The game</u> G <u>has no Nash equilibrium</u>.

This follows from either one of the inequality systems :

$$\{b_1 < a_1\ ,\ c_1 < d_1\ ;\ a_2 < c_2\ ,\ d_2 < b_2\}$$

or

$$\{a_1 < b_1\ ,\ d_1 < c_1\ ;\ c_2 < a_2\ ,\ b_2 < d_2\}$$

Then the mixed extension G_m has a <u>unique</u> NE outcome, namely a completely mixed equilibrium :

$$\mu_1^\star = \Bigg(\underbrace{\frac{d_2 - b_2}{a_2 + d_2 - b_2 - c_2}}_{\text{Top}} , \underbrace{\frac{a_2 - c_2}{a_2 + d_2 - b_2 - c_2}}_{\text{Bottom}} \Bigg)$$

$$\mu_2^\star = \Bigg(\underbrace{\frac{d_1 - c_1}{a_1 + d_1 - b_1 - c_1}}_{\text{Left}} , \underbrace{\frac{a_1 - b_1}{a_1 + d_1 - b_1 - c_1}}_{\text{Right}} \Bigg) \qquad (14)$$

and the corresponding equilibrium payoffs are :

$$v_1 = \frac{a_1 d_1 - b_1 c_1}{a_1 + d_1 - b_1 - c_1} \qquad v_2 = \frac{a_2 d_2 - b_2 c_2}{a_2 + d_2 - b_2 - c_2} \qquad (15)$$

The formula are a straightforward application of Lemma 3 and equation (13) : in view of the assumptions of Case 2, either U_i is a regular matrix or we can add a constant function c to U_i and make $U_i + c$ regular. This operation leaves intact the NE set and add the very constant c to the NE payoffs.

<u>Case 3</u> :

<u>The game</u> G <u>has two Nash equilibria</u>.

This follows from either one of the systems :

$\{b_1 < a_1 , c_1 < d_1 ; c_2 < a_2 , b_2 < d_2\}$

or

$\{a_1 < b_1 , d_1 < c_1 ; a_2 < c_2 , d_2 < b_2\}$

Then a new outcome arises in G_m , namely the completely mixed equilibrium $(\mu_1^\star , \mu_2^\star)$ given by (14) :

$$NE(G_m) = NE(G) \cup \{(\mu_1^\star , \mu_2^\star)\}$$

As a matter of exercize, the reader can check that Cases 1, 2, 3 exactly partition the family of 2 × 2 bimatrix games satisfying our one-to-one assumption.

Exercise 6.

Show an example of a 2 × 2 bimatrix game G where no player has equivalent pure strategies and the set $NE(G_m)$ is infinite.

Exercise 7. 2 × 2 zero-sum games.

Consider the family of 2 × 2 zero-sum games :

$$U_1 = \begin{bmatrix} a & c \\ b & d \end{bmatrix}$$

Prove that they are partitioned in two subsets.

Case 1 : The intersection $[a , d] \cap [b , c]$ is non empty.

Then G has a value and no new insight is gained by looking at its mixed extension.

Case 2 : The intersection $[a , d] \cap [b , c]$ is empty.

Then G has no value and G_m has a unique saddle-pair, actually a completely mixed saddle-pair. The mixed value of G is :

$$v_m(G) = \frac{a\,d - b\,c}{a + d - b - c}$$

Exercise 8.

If at least one player has two pure strategies, give a geometric method to compute the mixed value and optimal mixed strategies of any corresponding two-person zero-sum game. Application :

$$\begin{bmatrix} 0 & 3 & 2 & 5 & 3/2 \\ 6 & 2 & 1 & 0 & 2 \end{bmatrix}$$

Exercise 9.

Let (X_1 , X_2 , U_1) be a two-person zero-sum game of which the mixed extension has no completely mixed saddle-pair.

For all $(x_1 , x_2) \in X_1 \times X_2$ denote by $V_1 (x_1 , x_2)$ the mixed value of the game

$$(X_1 \setminus \{x_1\} , X_2 \setminus \{x_2\} , U_1)$$

namely the value of U_1 after row x_1 and column x_2 have been deleted.

Prove that the game

$$(X_1 , X_2 , V_1)$$

has a saddle-pair in pure strategies, and that its value is the mixed value of G.

Problem 4. The inspection game.

The game G_m is played in n steps by the customs officer (player 1) against the smuggler (player 2). At each step, player 2 shows up at the customs : player 1 decides to inspect him or not whereas player 2 decides to carry hot candies or not. Of course both decisions are made independently. Player 1 can inspect at each step (at some cost, see below) whereas player 2 will carry the hot stuff exactly once. If at step n player 1 inspects a candy-less player 2, he pays 1 to player 2 and G_{n-1} is played. If player 1 fails to inspect the hot luggage of player 2, player 2 receives c $(c \geqslant 1)$ and the game is over. If player 1 finds out the hot candies he receives a fine f $(f \geqslant c)$ and the game is over. If no inspection occurs and the luggage is not hot, then no payment occurs and G_{n-1} is played.

1) Find out an induction relation from v_{n-1}, the mixed value of G_{n-1}, to the mixed value v_n of G.

2) Compute v_n and the optimal mixed strategies of both players.

Problem 5.

Fix X_1, X_2, both finite, and for any $X_1 \times X_2$ matrix U_1 denote by $v_m(U_1)$ the mixed value of the (zero-sum) game (X_1, X_2, U_1).

1) Prove that $v_m(U_1)$ is a continuous function of U_1.

2) Prove the existence of an open dense subset Ω of $\mathbb{R}^{X_1 \times X_2}$ such that for all U_1 in Ω, the mixed extension of game (X_1, X_2, U_1) has a unique saddle-pair.

Problem 6. Profitable utility losses at a completely mixed Nash equilibrium. (Moulin [1976]).

Let $G = (X_1, X_2, U_1, U_2)$ be a bimatrix game where X_1 and X_2 have the same cardinality and U_1, U_2 are two regular $X_1 \times X_2$ matrices. Throughout the problem we assume that G_m has a (necessarily unique) completely mixed equilibrium (μ_1, μ_2) given by Lemma 3.

We set now : $\nu_1 = v_1 \cdot \mathbb{1}_{X_2} U_1^{-1}$. Thus ν_1 is a (row) vector in $\mathbb{R}^{X_1}$ and its components have an arbitrary sign. Notice however that $\nu_1 \cdot \mathbb{1}_{X_1} = 1$ so that some components of ν_1 are non-negative.

1) Prove that ν_1 is a mixed strategy of player 1 if and only if the completely mixed NE payoff v_1 (given by (12)) is the mixed value of the zero-sum game (X_1 , X_2 , U_1) :

$$\nu_1 \in M_1 \Leftrightarrow v_1 = v_m(U_1)$$

If ν_1 is indeed in M_1 then the completely mixed NE payoff of player 1 is just his secure utility of level ; further ν_1 is a prudent mixed strategy of player 1 and μ_1 is not (in general) prudent.

2) Suppose that ν_1 is not a mixed strategy (which by question 1 is equivalent to : $v_1 > v_m(U_1)$) then the set Y_1 of strictly negative components of ν_1 is non empty.

We say that a $X_1 \times X_2$ matrix L_1 is a Y_1-loss matrix if we have :

$$\begin{cases} \forall\, x_1 \in Y_1 \quad \forall\, x_2 , y_2 \in X_2 : L_1\ (x_1 , x_2) = L_1\ (x_1 , y_2) \geqslant 0 \\ \forall\, x_1 \in X_1 \setminus Y_1 \quad \forall\, x_2 \in X_2 : L_1\ (x_1 , x_2) = 0 \end{cases}$$

Show that for any small enough, non-zero Y_1-loss matrix L_1 , the game $(X_1 , X_2 , U_1 - L_1 , U_2)$ has a unique completely mixed equilibrium and the corresponding payoff vector $(\tilde{v}_1 , \tilde{v}_2)$ is such that :

$$\begin{cases} v_1 < \tilde{v}_1 \\ v_2 = \tilde{v}_2 \end{cases}$$

Interpretation.

Player 1 suffers a loss $L(x_1, .)$ whenever he uses a strategy $x_1 \in Y_1$. This penalty actually improves upon his own completely mixed NE payoff whereas player 2's payoff remains unchanged.

3) Consider a 2 × 2 game where both payoff functions are one-to-one (as in Example 3 above). Characterize those games for which the situation of question 2 arises. Give a numerical example.

III. INFINITE GAMES

When the sets of pure strategies X_i are infinite, even two-person zero-sum games may fail to have a value when the players use randomized strategies. A fortiori the existence of a mixed NE outcome is not guaranteed.

Example 4. Chinese poker.

Each one of the two players pick a (non-negative) integer. The player who called the largest number wins a dollar :

$$X_1 = X_2 = \mathbb{N} \qquad u_1(x_1, x_2) = \begin{cases} +1 & \text{if } x_2 < x_1 \\ 0 & \text{if } x_2 = x_1 \\ -1 & \text{if } x_1 < x_2 \end{cases}$$

A probability distribution over $X_i = \mathbb{N}$ takes the form :

$$\mu_i = (\mu_i(n))_{n\in\mathbb{N}} \quad \text{where} \quad \sum_{n\in\mathbb{N}} \mu_i(n) = 1 \,,\; \mu_i(n) \geqslant 0 \,,\; \text{all } n \in \mathbb{N}$$

Denoting by M_i the set of such distributions we face the mixed game $(M_1 , M_2 , \bar{u}_1)$ where :

$$\bar{u}_1 (\mu_1 , \mu_2) = \sum_{\substack{x_1 , x_2\in\mathbb{N} \\ x_2 < x_1}} \mu_1(x_1) \,.\, \mu_2(x_2) - \sum_{\substack{x_1 , x_2\in\mathbb{N} \\ x_1 < x_2}} \mu_1(x_1) \,.\, \mu_2(x_2)$$

The initial game is a zero-sum game without a value :

$$\sup_{x_1} \inf_{x_2} u_1 = -1 \qquad < +1 = \inf_{x_2} \sup_{x_1} u_1$$

Actually using mixed strategies does not improve <u>at all</u> the secure utility levels of any player, so that the mixed game $(M_1 , M_2 , \bar{u}_1)$ has no value as well :

$$\sup_{\mu_1} \inf_{\mu_2} \bar{u}_1 = -1 \qquad < +1 = \inf_{\mu_2} \sup_{\mu_1} \bar{u}_2 \tag{16}$$

To check this inequality, fix a probability distribution μ_1. Then for all $\varepsilon > 0$, there exists an integer n_ε such that :

$$\sum_{n=n_\varepsilon}^{+\infty} \mu_1(n) < \varepsilon$$

Next consider the (pure) strategy $x_2 = n_\varepsilon$ of player 2 :

$$\bar{u}_1 (\mu_1 , \delta_{x_2}) = - \sum_{n<n_\varepsilon} \mu_1(n) + \sum_{n>n_\varepsilon} \mu_1(n) < - (1 - \varepsilon) + \varepsilon = - 1 + 2\,\varepsilon$$

Since ε was arbitrary we deduce :

$$\inf_{\mu_2} \bar{u}_1 (\mu_1 , \mu_2) = \inf_{x_2} \bar{u}_1 (\mu_1 , \delta_{x_2}) = -1$$

implying the left-hand side equality in (16). As our game is symmetrical (see Exercise 1), the right-hand side equality follows.

The difficulty in the above example is the lack of compactness of the pure strategy sets as well as of the mixed strategy sets. When these two are compact, continuity of the (initial) payoff function is enough to guarantee the existence of a mixed Nash equilibrium.

Definition 3.

Let $G = (X_i , u_i ; i \in N)$, the initial N-normal form game, be such that :

$$\left.\begin{array}{l} X_i \text{ is a compact subset of some} \\ \quad \text{euclidian space } \mathbb{R}^{p_i} \\ u_i \text{ is a continuous function on } X_N \end{array}\right\} \text{ for all } i \in N$$

A mixed strategy μ_i of player i is a Radon probability measure on X_i : it is a positive continuous linear form on the set of continuous real functions on X_i endowed with the uniform convergence topology ; its value on the 1-constant function is 1. We denote by M_i the set of player i's mixed strategies.

The mixed extension of G is the game

$G_m = (M_i , \bar{u}_i ; i \in N)$:

$$\begin{cases} \bar{u}_i(\mu) = \displaystyle\int_{X_N} u_i(x) \, d\,\mu(x) \\ \text{where } d\,\mu(x) = \bigotimes_{i \in N} d\,\mu_i(x_i) \text{ is the product of the probability measures } \mu_i , \ i \in N \end{cases}$$

Theorem 3. (Glicksberg [1952]).

Under the assumptions of Definition 3, the mixed game G_m has at least one Nash equilibrium outcome.

Proof.

As a standard result of functional analysis (see e.g. Dunford and Schwartz [1957]) we have :

M_i is weakly compact within the dual space of the space of continuous functions on X_i , endowed with the uniform convergence topology.

$\bar{u}_i$ is linear w.r.t. its variable μ_i and continuous over M_N when the latter is endowed with the product of weak topologies.

Thus Glicksberg's theorem follows from Nash's (Theorem 2, Chapter III).

∎

Example 5. A location game.

Two shop owners must decide the location of their respective shops along the [0 , 1] interval. Player 1 supplies cheap sport equipment whereas player 2 supplies elegant sport suites. As side-by-side comparison is on average favourable to the cheap supply, and the players face an inelastic demand, player 1 wants to be located as close as possible to player 2, whereas player 2 tries to move away as far as possible from player 1. More specifically we assume that the profit functions take the following form :

$$\begin{cases} X_1 = X_2 = [0 , 1] \\ u_1 (x_1 , x_2) = 1 - |x_1 - x_2| \\ u_2 (x_1 , x_2) = |x_1 - x_2| \quad \text{if } |x_1 - x_2| \leq \frac{2}{3} \\ \qquad\qquad\quad = \frac{2}{3} \qquad\qquad \text{if } |x_1 - x_2| \geq \frac{2}{3} \end{cases}$$

Notice that the disexternalities imposed by player 1 on player 2 vanish when their distance is at least 2 / 3.

In the initial game no Nash equilibrium exists. This is illustrated on Figure 1 where the best response correspondences of both players are drawn.

Figure 1

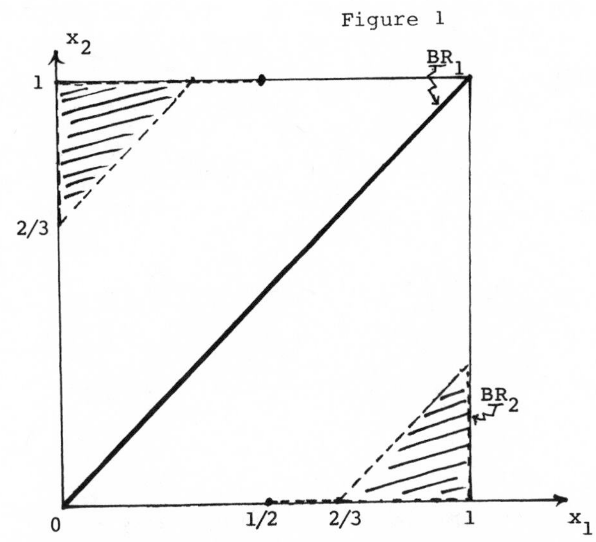

Since X_1 are compact and u_i are continuous, Theorem 3 shows the existence of at least one mixed NE outcome.

The following pair $(\mu_1^\star, \mu_2^\star)$ of mixed strategies does the job :

$$\begin{cases} \mu_1^\star = \frac{1}{3}\delta_0 + \frac{1}{6}\delta_{\frac{1}{3}} + \frac{1}{6}\delta_{\frac{2}{3}} + \frac{1}{3}\delta_1 \\ \mu_2^\star = \frac{1}{2}\delta_0 + \frac{1}{2}\delta_1 \end{cases}$$

Neither of these two mixed strategies are completely mixed (Definition 2 generalizes straightforwardly through the concept of carrier for a Radon measure : see e.g. Rudin [1966]). However $(\mu_1^\star, \mu_2^\star)$ share the typical property of completely mixed equilibria, namely :

$$\bar{u}_1 (\mu_1 , \mu_2^\star) = \bar{u}_1 (\mu_1^\star , \mu_2^\star) \quad \text{all } \mu_1 \in M_1$$

$$\bar{u}_2 (\mu_1^\star , \mu_2) = \bar{u}_2 (\mu_1^\star , \mu_2^\star) \quad \text{all } \mu_2 \in M_2 \qquad (17)$$

To check the first of these properties we fix a <u>pure</u> strategy x_1 of player 1 and we compute :

$$\bar{u}_1 (\delta_{x_1} , \mu_2^\star) = \frac{1}{2} (1 - x_1) + \frac{1}{2} (1 - (1 - x_1)) = \frac{1}{2} , \text{ all } x_1 \in X_1$$

As in the finite case, if $\bar{u}_1 (. , \mu_2^\star)$ is constant over pure strategies,

it is constant as well over mixed strategies (since the latter are limits of convex combinations of pure strategies).

We fix next a pure strategy x_2 of player 2 and we compute :

$$\bar{u}_2 (\mu_1^\star , \delta_{x_2}) = \frac{1}{3} x_2 + \frac{1}{6} \frac{1}{3} (\frac{1}{3} - x_2) + \frac{1}{6} \frac{2}{3} (\frac{2}{3} - x_2) + \frac{1}{3} . \frac{2}{3} \quad \text{if } 0 \leqslant x_2 \leqslant \frac{1}{3}$$

$$= \frac{1}{3} x_2 + \frac{1}{6} (x_2 - \frac{1}{3}) + \frac{1}{6} \frac{2}{3} (\frac{2}{3} - x_2) + \frac{1}{3} (1 - x_2) \quad \text{if } \frac{1}{3} \leqslant x_2 \leqslant \frac{2}{3}$$

$$= \frac{1}{3} . \frac{2}{3} + \frac{1}{6} (x_2 - \frac{1}{3}) + \frac{1}{6} (x_2 - \frac{2}{3}) + \frac{1}{3} (1 - x_2) \quad \text{if } \frac{2}{3} \leqslant x_2 \leqslant 1$$

henceforth :

$$\bar{u}_2 (\mu_1^\star , \delta_{x_2}) = \bar{u}_2 (\mu_1^\star , \mu_2) = \frac{7}{18} \quad \text{all } x_2 \in X_2$$

$$\text{all } \mu_2 \in M_2$$

Remark that μ_2^* is an optimal prudent strategy for player 1 in the zero-sum game $(M_1, M_2, \bar{u}_1)$. Namely, $\bar{u}_1(\mu_1, \mu_2) = \bar{u}_1(\mu_2, \mu_1)$ for all μ_1, μ_2, hence the top equation in (17) implies that (μ_2^*, μ_2^*) is a saddle-pair of $\bar{u}_1$. Thus the mixed value of u_1 is $\frac{1}{2}$ and coincides with player 1's NE payoff.

A similar argument shows that (μ_1^*, μ_1^*) is the (unique) saddle-pair of game $(M_1, M_2, -\bar{u}_2)$, so the (mixed) secure utility of player 2 is $\frac{7}{18}$ and equals his NE payoff.

Observe finally that our NE outcome is Pareto dominated.

The computation of the mixed NE outcomes is very difficult to perform even when the initial strategy sets are convex compact subsets of euclidian spaces. Hence we shall not explore this issue and refer the reader to Parthasarathy-Raghavan [1971], Tijs [1981] and the bibliography there.

Our two next examples are two games where each player strategy set is one dimensional and the payoff function is discontinuous. In the first of these games, the payoffs are uniformly bounded, yet <u>no mixed Nash equilibrium</u> exists, therefore stressing the role of the payoff's continuity in Glicksberg theorem.

Example 6. A two-person zero-sum game on the unit square with no mixed value, (Sion and Wolfe [1957]).

We set $X_1 = X_2 = [0, 1]$ and :

$$u_1(x_1, x_2) = -1 \quad \text{if} \quad x_1 < x_2 < x_1 + \frac{1}{2}$$

$$= 0 \quad \text{if} \quad x_1 = x_2 \text{ or } x_2 = x_1 + \frac{1}{2}$$

$$= +1 \quad \text{if} \quad x_2 < x_1 \text{ or } x_1 + \frac{1}{2} < x_2$$

We claim that the mixed extension of this two-person zero-sum game has no value. More precisely we have :

$$\sup_{\mu_1} \inf_{\mu_2} \bar{u}_1 = \frac{1}{3} < \frac{3}{7} = \inf_{\mu_2} \sup_{\mu_1} \bar{u}_1$$

We prove the right-hand side equation by deriving a contradiction from :

$$\sup_{x_1} \bar{u}_1 (\delta_{x_1}, \mu_2) < \frac{3}{7} \tag{18}$$

Namely taking successively $x_1 = 1$, $x_1 = 0$, (18) implies :

$$\mu_2(\{1\}) > \frac{4}{7} \tag{19}$$

$$\mu_2(\{1\}) + b - a < \frac{3}{7} \tag{20}$$

where we set $a = \mu_2 \left(]0, \frac{1}{2}[\right)$ $b = \mu_2 \left(]\frac{1}{2}, 1[\right)$. Next we apply

(18) at $x_1 = \frac{1}{2} - \varepsilon$ and we let ε go to zero :

$$\mu_2(\{0\}) + \mu_2(\{1\}) - \mu_2(\{\tfrac{1}{2}\}) + a - b < \frac{3}{7} \tag{21}$$

Summing up (20) and (21) we get :

$$2\,\mu_2(\{1\}) - \mu_2(\{\tfrac{1}{2}\}) < \frac{6}{7} \tag{22}$$

On the other hand, μ_2 is a probability, hence :

$$\mu_2(\{1\}) + \mu_2(\{\tfrac{1}{2}\}) + a \leqslant 1 \tag{23}$$

Summing up (23) and (20) we get :

$$2\,\mu_2(\{1\}) + \mu_2(\{\tfrac{1}{2}\}) < \frac{10}{7} \tag{24}$$

Summing up (24) and (22) yields a contradiction of (19). Thus we have proved

$$\inf_{\mu_2} \sup_{\mu_1} \bar{u}_1 \geqslant \frac{3}{7}$$

Now taking a mixed strategy $\mu_2^\star$ such as :

$$\mu_2^\star = \frac{1}{7}\,\delta_{\frac{1}{4}} + \frac{2}{7}\,\delta_{\frac{1}{2}} + \frac{4}{7}\,\delta_1$$

clearly guarantees a maximal loss $\frac{3}{7}$ to player 2 :

$$\sup_{\mu_1} \bar{u}_1(\mu_1, \mu_2^\star) = \frac{3}{7}$$

Exercise 10.

Prove that $\sup_{\mu_1} \inf_{\mu_2} \bar{u}_1 = \frac{1}{3}$

<u>Hint</u> : A prudent strategy of player 1 is

$$\mu_1^\star = \frac{1}{3}\delta_0 + \frac{1}{3}\delta_{\frac{1}{2}} + \frac{1}{3}\delta_1$$

In our last example, the strategy set are real half-lines and the pay-off function are discontinuous and unbounded. Yet a nice completely mixed NE outcome exists.

<u>Example 7</u>. <u>The auction dollar game</u>.

A dollar is allocated by a sealed bid auction among the n agents. Say that the various bids are $x_1, \ldots, x_n$ and the highest bid is that of player i_o : $w(x) = \{i_o\}$ where $w(x)$ is the set of winning players at x, namely :

$$w(x) = \{ i \in N / x_i = \sup_{j \in N} x_j \}$$

The rule is that i_o wins the dollar at the second announced price and <u>every other bid is paid as well</u> : each player i, $i \neq i_o$, must pay his actual bid x_i.

Assuming that ties are broken by flipping a coin among the winning players to allocate the dollar, we obtain the following normal form game :

$$\left.\begin{aligned} X_i &= [0, +\infty[\\ u_i(x) &= 1 - \sup_{j \in N \setminus \{i\}} x_j && \text{if } w(x) = \{i\} \\ &= \frac{1}{|w(x)|} - x_i && \text{if } i \in w(x) \text{ and } |w(x)| \geq 2 \\ &= -x_i && \text{if } i \notin w(x) \end{aligned}\right\} \quad (25)$$

This game is a variant of the war of attrition (Problem 5, Chapter III) where the object has the same value to every player, and the number of players is arbitrary. The analysis of the initial normal form game (25) parallels that of the war of attrition. The game (25) has essentially n distinct pure strategy N E outcomes, all of them Pareto optimal, where one player bids above 1 whereas the other players bid nothing.

Each of these equilibria involves a sharp dissymetry among the players : to make such an equilibrium the plausible outcome of our game, a player must act as a leader by committing himself to bid above 1. If no such commitment is possible, or equivalently if the players can not be discriminated according to their committment ability, (a likely assumption if the players are aware that the game is symmetrical) an outcome where some player is leader lacks realism.

Mixed strategies, which require players to actually pick their pure strategies secretly, quite on the opposite of a leadership behaviour, raise here a symmetrical and completely mixed Nash equilibrium.

To prove this claim we exhibit a probability distribution ν^* on $[0,+\infty[$ such that for all $i \in N$:

$$\bar{u}_i(\delta_x, u_{\hat{i}}^*) \text{ does not depend on } x \in X_i \text{ given that } u_j^* = \nu^*, \text{ all } j \in N \setminus \{i\} \qquad (26)$$

The probability distribution ν^* is taken to have a continuous density function f :

$$\nu^*(A) = \int_A f(t)\,dt \quad \text{all measurable } A \subset [0,+\infty[$$

Denote by F the cumulative distribution :

$$F(x) = \int_0^x f(t)\,dt$$

When all players j of $N \setminus \{i\}$ play the mixed strategy ν^*, the random variable

$$\sup_{j \in N \setminus \{i\}} x_j$$

has the cumulative distribution F^{n-1}, with corresponding density $f_n = (n-1)F^{n-2}f$. Therefore we can compute :

$$\bar{u}_i(\delta_x, u_{\hat{i}}^*) = \int_0^x (1-t)\,f_n(t)\,dt - x\int_x^{+\infty} f_n(t)\,dt =$$

$$= (n-1)\int_0^x (1-t)\,F^{n-2}(t)\,f(t)\,dt - x\,(1 - F^{n-1}(x))$$

In view of our continuity assumption on f, the above expression is differentiable w.r.t. x and condition (26) amounts to :

$$0 = \frac{\partial \bar{u}_i}{\partial x}(\delta_x, u_{\hat{1}}^{\star}) = (n-1)\, F^{n-2}(x)\, f(x) + F^{n-1}(x) - 1$$

Hence $G = F^{n-1}$ is solution of $G'(x) + G(x) = 1$ and taking $F(0) = 0$ into account, we get finally :

$$F(x) = [\, 1 - e^{-x}\,]^{\frac{1}{n-1}} \qquad \text{all } x \geqslant 0$$

Thus the symmetrical completely mixed NE outcome has each player using the same density function :

$$f(x) = \frac{1}{n-1} \cdot \frac{e^{-x}}{[\, 1-e^{-x}\,]^{\frac{n-2}{n-1}}}$$

Notice that for an arbitrary large number λ, the probability that a player bids above λ is non zero (it is $1 - (1-e^{-\lambda})^{\frac{1}{n-1}}$), thus implying that our NE outcome is Pareto dominated. However the expected value α of the winning bid is finite :

$$\alpha = \frac{n}{n-1} \int_0^{\infty} [\, 1 - e^{-t}\,]^{\frac{1}{n-1}}\, e^{-t}\, t\, dt$$

Remark finally that the NE pay-off to every player is exactly zero, as an easy computation shows.

Problem 6. A modified chinese poker with a value.

The two players pick an integer. If these two numbers differ the pay-off is zero. If they coïncide $x_1 = x_2 = p$, player 1' pay-off is a_p where $a_1 \leqslant a_2 \leqslant \ldots \leqslant a_p \leqslant \ldots$ is a non decreasing sequence of positive numbers. Thus player 2 faces a dilemma.

The larger number that he picks, the less probable it is that player 1 can guess it, but the more risky a potential discovery.

The initial two-person zero zum game is $G = (X_1, X_2, u_1)$:

$$X_1 = X_2 = \mathbb{N}$$

$$\begin{aligned} u_1(x_1, x_2) &= a_p \quad \text{if } x_1 = x_2 = p \\ &= 0 \quad \text{if } x_1 \neq x_2 \end{aligned}$$

Its mixed extension can be defined as :

$$M_1 = M_2 = M = \{ \nu \in \mathbb{R}^{\mathbb{N}} / \sum_{p=1}^{+\infty} \nu(p) = 1 \text{ and } \nu(p) \geqslant 0, \text{all } p \}$$

$$\bar{u}_1(\mu_1, \mu_2) = \sum_{p=1}^{+\infty} a_p \, \mu_1(p) . \mu_2(p)$$

with the qualification that $\bar{u}_1$ can take the value $+\infty$. Alternatively we can define :

$$\tilde{M}_1 = \tilde{M}_2 = \{ \nu \in M / \sup_{p \in \mathbb{N}} [a_p . \nu(p)] < +\infty \}$$

so that $\bar{u}_1$ is always finite valued. Either one of these two models can be used in the subsequent questions :

1) Prove that if :

$$\sum_{p=1}^{+\infty} \frac{1}{a_p} < + \infty$$

the mixed extension of G has a value, each player has a unique optimal strategy $\nu^{\star}$, and $\nu^{\star}$ is completely mixed.

2) Prove that if :

$$\sum_{p=1}^{+\infty} \frac{1}{a_p} = + \infty$$

the mixed extension G has a value but no player has an optimal strategy.

REFERENCES

DUNFORD, N. and J.T. SCHWARTZ. 1957. Linear operators. New York: John Wiley and Sons.

GLICKSBERG, I.L. 1952."A further generalization of the Kakutani fixed point theorem with application to Nash equilibrium points". Journal American Society 3, 170-174.

GUILBAUD, G. 1968. Elements de la théorie mathématique des jeux. Paris : Dunod

LUCE, R.D. and H. RAIFFA. 1957. Games and decisions. New York : J. Wiley and Sons.

MOULIN, H. 1975. "Extension of two-person zero-sum games". Journal of Mathematical Analysis and Applications 55, 2 : 490-507.

MOULIN, H. 1981. Théorie des jeux pour l'économie et la politique. Paris : Hermann ed.

PARTHASARATHY, T. and T.E.S. RAGHAVAN . 1971. Some topics in two-person games. New York : American Elsevier.

RIVES, N.W. 1975. "On the history of the mathematical theory of games". Hope 7, 4.

RUDIN, W. 1966. Real and complex analysis, New York : Mc Graw Hill .

SION, M. and P. WOLFE. 1957. "On a game without a value". Contributions to the theory of games, vol.3. Annals of Maths. Studies 39. Princeton : Princeton University Press.

TIJS, S.H. 1981. "Nash equilibria for non-cooperative n-person games in normal form". SIAM Journal Appl. Math. 23, 2 : 225-237.

PART II : COOPERATIVE PLAYERS

Throughout part 1 we dealt with a society which allowed no explicit communication among players, as a result of physical, legal barriers and/or mutual distrust. Mutual observations of tactical moves was the only, indirect, way of conveying some strategical information (see the Cournot tatonnement Chapter III, Section 3).

On the contrary, we seek now to design behavioural patterns of a cooperative society, where explicit communication takes place. Structural inefficiency of the non-cooperative equilibrium outcome (an unambiguous feature in most of our non-zero sum examples in Part 1) is now interpreted as an incentive to cooperation. We attempt to describe its likely manifestations.

After cooperative communication, that may involve mutual disclosure of the utility functions, bargaining, and various psychological manoeuvres, the players reach a cooperative agreement. Agreements can be binding (when several players can sign a contract to use specified strategies, and this contract can be enforced by some judging authority to whom obedience is due by all players) or non binding (when no such authority exists so that an agreement, very much like an international treaty, will be respected so long as betrayal "does not pay").

Our approach consists of looking at reasonable outputs of the communication stage without attempting any description of this phase. Non binding agreements generate several stability questions that are best-discussed according to the informational set-up (Chapter V and VI). Binding agreements, on the other hand, raise the normative issue of equity, viewed as a powerful encouragement to cooperation. They will not be considered here (see Introduction) .

CHAPTER V. SELF-ENFORCING AGREEMENTS.

A non binding agreement specifies an outcome, that is a particular strategy for each player. Since we want each player to keep full sovereignty over his or her strategic variable, no player nor any coalition of players can actually be forced to use the recommended agreement strategy. Thus the only way to prevent a potential betrayal by an individual agent or a coalition of agents is to make this deviation non profitable to the deviants. This is the self-enforcing property.

Self-enforcement is much less easy a concept than it looks at first sight. Indeed a deviation by some players betraying a considered agreement might induce the other (non originally betraying) players to change their strategies as well. Such induced changes are hardly predictible, whether or not we make the assumption that initial betrayal has killed the cooperative spirit and drove back the players to a non-cooperative world. Namely the only undebatable conclusion of Part 1 is that in most normal form games several distinct non-cooperative equilibrium outcomes are on stage (see Chapter III and IV).

Accordingly a non-binding agreement will consist of an agreed upon outcome and a reaction scenario by which each player i announces his specific reaction to any deviation by the coalitions not containing i . In the next Chapter, these reaction scenarios will be threats, that can oppose a different reaction to each conceivable deviation from the agreement. In the current Chapter we look at a very particular reaction scenario, namely no reaction : loyalty to the agreement requires simply not to abandon the agreed upon strategy whatever strategies may be used by the rest of the world. In that context a stable agreement is called self-enforcing.

The fundamental example of a self-enforcing agreement is that of a Nash equilibrium, of which the stability is enhanced by mutual secrecy on the final strategic decision. Here we explore two variations of the NE concept. If the players can form betraying coalitions the stability of the agreed outcome is threatened by potential deviations from any coalition : this leads to the strong equilibrium concept (Section 1). If our players can randomly correlate their strategies and send a decentralized signal to each individual, then new self-enforcing agreements emerge that we call the correlated equilibrium outcomes (Section 2).

I. STRONG EQUILIBRIUM

Example 1. Two crowded ships : an externality game.

Two ships leave the same day to the traesure island and each one of our n pirates must make the decision to board on ship A or ship B. If t denotes the number of pirates on ship A then the journey on A will last $a(t)$ days, whereas the journey on ship B where $(n-t)$ pirates are boarded last for $b(n-t)$ days. As each player seeks only to minimize the length of his own journey, the situation is described by the following game :

$$\begin{cases} X_i = \{0, 1\} & i \in N \\ u_i(x) = -a(t) & \text{if } x_i = 1 \\ \quad\quad = -b(n-t) & \text{if } x_i = 0 \end{cases} \quad \text{where } t = \sum_{j \in N} x_j$$

where $x_i = 1$ stands for player i boarding on ship A.

We assume that a and b are both strictly increasing functions on $\{0,\ldots,n\}$ with $a(0) < b(n)$ and $b(0) < a(n)$.

A non cooperative equilibrium (N E outcome) of this game is any x^* such that $t^* = \sum_{i \in N} x_i^*$ satisfies :

$a(t^*) \leqslant b(n-t^*+1)$: no incentive to switch from strategy 1 to strategy 0

and

$b(n-t^*) \leqslant a(t^*+1)$: no incentive to switch from strategy 0 to strategy 1

Assuming for simplicity that $a(t) \neq b(n-t+1)$ for all t, there is exactly one integer t^* satisfying the above two inequalities, namely :

$$t^* = \sup\{t/a(t) \leqslant b(n-t+1)\} = \inf\{t/b(n-t) \leqslant a(t+1)\}$$

Thus an outcome x^* is a Nash equilibrium if and only if $\sum_{i \in N} x_i^* = t^*$.

If the players can not communicate before picking a strategy (they must reserve a seat on either one of the two ships before they ever meet) they can not coordinate their choices so as to reach a Nash equilibrium in pure strategy. Presumably they will play the symmetrical completely mixed equilibrium, which actually is Pareto dominated by anyone of the pure NE outcomes.

<u>Exercise 1</u>. Prove the existence of a unique completely mixed NE outcome where each pirate plays strategy 1 with the same probability p. Prove that it is in general Pareto dominated by any NE outcome in pure strategies.

Some coordination device here allows the players to agree on a particular NE outcome in pure strategies x^* : this agreement is not only robust against potential individual deviations (this is the NE property) but also against <u>coalitional</u> deviations. Namely suppose that for some coalition T, $T \subset N$, all agents i in T find it profitable to switch from strategy x_i^* to strategy $x_i = 1 - x_i^*$.

$$\begin{cases} u_i(x_T, x_{T^C}^*) \geq u_i(x^*) & \text{all } i \in T \\ u_i(x_T, x_{T^C}^*) > u_i(x^*) & \text{for at least one } i \in T \end{cases} \quad (1)$$

Suppose first $t = \sum_{i \in T} x_i + \sum_{i \in T^C} x_i^*$ equals t^*. Then at least one pair $j_1, j_2 \in T$ is such that :

$$\begin{cases} x_{j_1}^* = 0 & x_{j_1} = 1 \\ x_{j_2}^* = 1 & x_{j_2} = 0 \end{cases}$$

Applying (1) to j_1, j_2 successively yields :

$$\left.\begin{array}{l} a(t^*) \leq b(n-t^*) \\ b(n-t^*) \leq a(t^*) \end{array}\right\} \Rightarrow a(t^*) = b(n-t^*)$$

therefore $u_i(x_T, x_{T^C}^*) = u_i(x^*)$ all $i \in N$, a contradiction of (1).

Suppose now $t > t^*$. Then for at least one $i \in T$ we have :

$$x_i^\star = 0 \qquad x_i = 1$$

So that by (1) :

$$b(n - t^\star) \geqslant a(t)$$

Since $b(n - t^\star) < a(t^\star + 1)$ we get finally $t < t^\star + 1$ a contradiction.

We have proved that any agreement to play a particular NE outcome, is robust as well against coalitional deviations : it is a strong equilibrium outcome (Definition 1 below). If $a(t^\star)$ and $b(n - t^\star)$ are sufficiently close, each player essentially receives the same agreement pay-off so that the choice of one particular NE outcome is non conflictual. To actually implement one of these, the pirates successively and publicly pick a ship by the following algorithm.

$$x_i = 1 \quad \text{if} \quad a(x_1 + \ldots + x_{i-1} + 1) < b(i - [x_1 + \ldots + x_{i-1}])$$

$$x_i = 0 \quad \text{if} \quad b(i - [x_1 + \ldots + x_{i-1}]) < a(x_1 + \ldots + x_{i-1} + 1)$$

Definition 1.

Given a N-normal form game $G = (X_i, u_i\ ; i \in N)$, we say that $x^\star$ is a strong equilibrium outcome if no coalition of players can profitably deviate given that the complement coalition will not react :

$$\forall T \subset N \quad \forall x_T \in X_T \quad \text{No} \begin{cases} \forall i \in T & u_i(x_T, x^{\star}_{T^c}) \geq u_i(x^{\star}) \\ \exists i \in T & u_i(x_T, x^{\star}_{T^c}) > u_i(x^{\star}) \end{cases} \qquad (2)$$

We denote by $SE(G)$ the - possibly empty - set of strong equilibrium outcomes of G.

Taking $T = N$ in (2) yields that a SE outcome is a Pareto optimum. Taking $T = \{i\}, i \in N$ yields that it is a Nash equilibrium outcome as well. Clearly in two-person games a SE outcome simply is a Pareto optimal NE outcome.

We interpret the stability property (2) by a two-stage decision making process. In the first stage our players congregate to agree upon a particular outcome $x^{\star}$. Next communication is banned and each individual has to make a single isolated decision on his final irrevocable strategy. Each one can freely betray his word to play $x^{\star}_i$ but he cannot inform the non-betraying players of his own deviation. Also any coalition T can plot a jointly decided betrayal x_T, but again the players outside the conspiracy can not be informed of the intended change of strategy, and therefore should be expected to stick to the agreed upon behaviour $x^{\star}_{T^c}$. This limitation to communication is crucial to the plausibility of Definition 1 as our example 2 below illustrates "a contrario". See also Definition 1 Chapter III and comments thereafter (notice that we do not invoke any tatonnement process to enhance the strong equilibrium concept, since the myopic behavior can not be justified in a cooperative world).

We have seen that in two person games, the stability of any N E outcome is threatened if the struggle for the leadership ever occurs (see Example 2 and Lemma 2 Chapter III). A similar configuration is possible in n-players games, as illustrated by our next example.

<u>Example 2</u>. <u>A bargaining game</u>.

The n players must share a dollar. They submit individual claims that a referee satisfies if they are feasible altogether. Otherwise each player gets nothing :

$$\begin{cases} X_i = [0,1] & \text{all } i \in N \\ u_i(x) = x_i & \text{if } \sum_{j \in N} x_j \leq 1 \\ = 0 & \text{if } \sum_{j \in N} x_j > 1 \end{cases}$$

We assert that :

$$S\,E\,(G) = \{x \in X_N / \sum_{i \in N} x_i = 1\}$$

in other words an outcome is a strong equilibrium if and only if the corresponding claims add up to exactly one dollar. We let the reader check this assertion.

Observe that if coalition T, acting as a leader, commits itself to a particular strategy T-uple x_T^* with

$\sum_{i \in T} x_i^* = 1 - \varepsilon$ ($\varepsilon > 0$ is small), then coalition T^c acting as a follower optimally replies by a vector of claims x_{T^c} such that :

$$\sum_{i \in T^c} x_{T^c} = \varepsilon$$

Hence any individual or coalition, who is endowed with enough commitment power will nearly extract all the potential benefit of 1 dollar! Thus the struggle for the leadership is likely to be intense in this bargaining game.

Pick now a particular SE outcome x^* upon which the players may have agreed, say $x_i^* = \frac{1}{n}$, all $i \in N$. To self-enforce this agreement, each player must commit himself to never listen to any claim above $\frac{1}{n}$ by anyone threatening to take the leadership. This deafness policy is optimally achieved by a careful destruction of the communication channels among players, afterwhich every player can still change at will his or her strategy but can not do it publicly anymore.

In a game like Example 2, the crucial stability feature of the agreement to play a particular SE outcome, relies on the restriction to communication that the players must enforce legally (like in a secret ballot regulation) or physically (like Ulysses companions filling their ears of wax). Henceforth a non-binding agreement involves some binding limitation to communication. This point will be made quite transparent by the concept of correlated equilibria to which is devoted our next Section.

We will see in the next Chapter that a symmetrical informational constraint is required by the deterring purpose of threats (see Chapter VI).

We already know (Chapter III, Section 2) that a normal form game whose strategy sets are open subsets of euclidian spaces, and utility functions are differentiable, can not be expected in general to possess a Pareto optimal N E outcome. A fortiori it should not possess a S E outcome. Under very strong convexity assumptions, however, existence of a S E outcome can be guaranteed (see Aumann [1959]).

Exercise 2. Generalization of Example 1.

Each player must pick one among p local public goods. If t_k is the total number of players deciding for the good k, $k = 1,\ldots,p$, then every consumer of k enjoys the utility level $-a_k(t_k)$ where $a_k(.)$ is the disutility of consuming good k as a strictly increasing function of t_k.

Henceforth the following game :

$$\begin{cases} X_i = \{1,\ldots,p\} & \text{all } i \in N \\ u_i(x) = -a_k(t_k) & \text{if } x_i = k \text{ and } t_k = |\{j \in N / x_j = k\}| \end{cases}$$

Assume $a_k(0) = 0$ and $a_k(n) = 1$ for all $k = 1,\ldots,p$.

Prove the existence of at least one strong equilibrium outcome for this game and give conditions under which all SE outcomes yield the same distribution $(t_1, \ldots, t_p)$.

Exercise 3. Location game for complementary shops.

Two shop owners decide the location of their respective shops along the [0,1] interval. They supply complementary goods (say sport equipment and travel-agent services), implying a positive externality from either shop to the other (contrary to example 5, Chapter IV). Moreover player 1 inclines to be located as close to 0 as possible whereas player 2 seeks to be located as far as possible from 0.

Specifically they face the following game :

$$\begin{cases} X_1 = X_2 = [0,1] \\ u_1(x_1, x_2) = \alpha_1 x_1 - |x_1 - x_2| \\ u_2(x_1, x_2) = \alpha_2 (x_2 - 1) - |x_1 - x_2| \\ \text{where } \alpha_1 < 0 < \alpha_2 \end{cases}$$

Suppose that $|\alpha_i| \leq 1 \quad i = 1, 2.$

Prove that outcome (x_1, x_2) is a strong equilibrium of this game if and only if $x_1 = x_2$. Discuss the other cases.

II - CORRELATED EQUILIBRIUM

Mixed strategies raise new Nash equilibrium outcomes (every finite game has at least one mixed N E outcome - Theorem 1, Chapter IV). From a cooperative point of view, a mixed Nash equilibrium is a non binding agreement self-enforced by the mutual privacy of the lotteries that each individual player builds up to pick at random his final decision (i.e. a pure strategy).

Example 3. An altruistic crossing game.

We modify the pay-offs of the crossing game (Example 2, Chapter III) : given that he will himself stop, a player now prefers that his partner goes :

	Stop	Go
Stop	1 / 1	$1+\varepsilon$ / 2
Go	2 / $1+\varepsilon$	0 / 0

(This game is known in the litterature as the "battle of the sexes", see Luce and Raïffa [1957]).

In addition to the two pure N E outcomes, with respective pay-off vectors $(1+\varepsilon, 2)$ and $(2, 1+\varepsilon)$, this 2 x 2 game has a completely mixed equilibrium namely

$$\mu_1^{\star} = \mu_2^{\star} = \frac{1+\varepsilon}{2+\varepsilon}\,\text{stop} + \frac{1}{2+\varepsilon}\,\text{go}$$

with corresponding pay-off vector $(1+\frac{\varepsilon}{2+\varepsilon}, 1+\frac{\varepsilon}{2+\varepsilon})$.

As the mixed N E outcome is Pareto dominated (by any pure N E outcome) it can at least be defended by an equity argument : it treats similarly our two identical players.

However one can argue against the mixed N E outcomes where each player ends up with only his (mixed) secure utility level $1+\frac{\varepsilon}{2+\varepsilon}$ since he is not even guaranteed of this utility level because the mixed N E strategies are not prudent.

To check this claim, observe that the mixed value of game (X_1, X_2, u_1) is $1+\frac{\varepsilon}{2+\varepsilon}$ with unique mixed saddle-pair $(\mu_1^{\circ}, \mu_2^{\star})$ where

$$\mu_1^{\circ} = \frac{2}{2+\varepsilon}\,\text{stop} + \frac{\varepsilon}{2+\varepsilon}\,\text{go}$$

Moreover :

$$u_1(\mu_1^{\star}, \text{go}) = \frac{1-\varepsilon^2}{2+\varepsilon} < 1+\frac{\varepsilon}{2+\varepsilon} = u_1(\mu_1^{\star}, \mu_2^{\star}) = u_1(\mu_1^{\circ}, \text{go})$$

which makes clear that the mixed N E strategy $\mu_1^{\star}$ is more risky than the prudent one μ_1°.

The only strategic argument to support a mixed equilibrium is really its stability. If the players can help each other to protect their respective privacy, after a non binding agreement to play the completely mixed N E outcome, then the stability of this agreement is self-enforced, and therefore each player accurately predicts the other's behavior. An agreement to play the prudent mixed strategy would, to the contrary, be self-deceiving and induce a sequence of anticipated best replies, thus making the final outcome quite impredictible.

Here the trade-off is between an outcome (the mixed N E) that will safely emerge if any fellow player is as rational as I am, but involve some genuine risk if he plays foolishly, and another one (the pair of prudent mixed strategies) where the worst risk is as safe as possible, but will induce each rational player to a profitable unilateral change.

Back to our crossing game, we construct a random device more subtle than independent randomization of the strategies, which allows an equilibrium outcome Pareto superior to the completely mixed N E outcome.

Example 3 bis. Crossing game with red lights.

The players build a random signal, that shows (green, red) or (red, green) with equal probability. Their agreement is that whoever sees green goes whereas whoever sees red stops. This agreement is self-enforcing since for each drawing of the lottery, the corresponding agreed deterministic outcome is a Nash equilibrium. The resulting expected pay-off is $\frac{3}{2} + \frac{\varepsilon}{2}$ to each player, henceforth Pareto optimal and fair.

Definition 2. (Aumann [1974]).

Given a game $G = (X_i, u_i, i \in N)$ with finite strategy sets, we denote by $L = (L(x))_{x \in X_N}$ a correlated lottery, that is a probability distribution over X_N. For all $i \in N$, and all $x_i \in X_i$ we denote by L_{x_i} the conditional probability of L over $X_{N \setminus \{i\}}$:

$$\text{all } x_{\hat{i}},\ L_{x_i}(x_{\hat{i}}) = \frac{1}{\sum_{y_{\hat{i}} \in X_{N \setminus \{i\}}} L(x_i, y_{\hat{i}})} \cdot L(x_i, x_{\hat{i}}) \text{ if the denominator is non zero.}$$

$$= 0 \text{ if } L(x_i, y_{\hat{i}}) = 0 \text{ for all } y_{\hat{i}} \in X_{N \setminus \{i\}}$$

We say that L is a correlated equilibrium of G if the following inequalities hold :

$$\forall i \in N\ \forall x_i, y_i \in X_i \sum_{x_{\hat{i}} \in X_{N \setminus \{i\}}} u_i(x_i, x_{\hat{i}}) L_{x_i}(x_{\hat{i}}) \geq \sum_{x_{\hat{i}} \in X_{N \setminus \{i\}}} u_i(y_i, x_{\hat{i}}) L_{x_i}(x_{\hat{i}}) \quad (3)$$

We denote by $CE(G)$ the set of correlated equilibria of G.

The cooperative scenario justifying Definition 2 is the following. The players cooperatively construct a random signal that potentially select a outcome $x \in X_N$ with probability $L(x)$. If outcome x is drawn, player i is informed of the component x_i only ; then each player selects freely independently and secretly his or her actual strategy. The signal x_i to player i is understood as a non-binding suggestion to play x_i. Condition (3) express that the agreement to obey this suggestion is self-enforcing, given the limited information available to each player. Namely suppose agent i is told by the signal to use strategy x_i. He infers from the overall distribution L that the $N \setminus \{i\}$-uple $x_{\hat{i}}$ has been selected with probability $L_{x_i}(x_{\hat{i}})$. Therefore :

$$[u_i(y_i, .), L_{x_i}] = \sum_{x_{\hat{i}} \in X_{N\setminus\{i\}}} u_i(y_i, x_{\hat{i}}) L_{x_i}(x_{\hat{i}})$$

is the expected utility of playing strategy $y_i \in X_i$ given that all other players obey the signal. Thus condition (3) expresses that using the strategy suggested by the signal is always optimal to player i given the information available to him and the assumption that the other players obey the signal.

Suppose that x_i is such that $L(x_i, x_{\hat{i}}) = 0$ for all $x_{\hat{i}} \in X_{N\setminus\{i\}}$ i.e. the probability that x_i is ever suggested by the signal is zero. For such an x_i condition (3) is trivially satisfied, hence we rewrite equivalently system (3) as :

$$\forall i \in N \ \forall x_i, y_i \in X_i \sum_{x_{\hat{i}} \in X_{N \setminus \{i\}}} u_i(x_i, x_{\hat{i}}) L(x_i, x_{\hat{i}}) \geq \sum_{x_{\hat{i}} \in X_{N \setminus \{i\}}} u_i(y_i, x_{\hat{i}}) L(x_i, x_{\hat{i}}) \quad (4)$$

From this it follows that a lottery L is a correlated equilibrium if and only if it satisfies a system of linear inequalities. Actually this system always has a solution as the next result demonstrates.

Lemma 1.

Notations and assumptions as in Definition 2. The set $CE(G)$ of correlated equilibriums of G is a non-empty convex compact subset of the unit simplex of $\mathbb{R}^{X_N}$. If $\mu = (\mu_i)_{i \in N}$ is an outcome of the mixed game G_m then the associated lottery $L = \bigotimes_{i \in N} \mu_i$

$$L(x) = \prod_{i \in N} \mu_i(x_i) \qquad (5)$$

is a correlated equilibrium of G if and only if μ is a Nash equilibrium of G_m.

Proof.

Let $\mu = (\mu_i)_{i \in N}$ be an outcome of the mixed game G_m and L be the corresponding product lottery given by (5). System (4) becomes :

$$\mu_i(x_i) \cdot \bar{u}_i(\delta_{x_i}, \mu_{\hat{i}}) \geq \mu_i(x_i) \cdot \bar{u}_i(\delta_{y_i}, \mu_{\hat{i}}) \text{ all } i, \text{ all } x_i, y_i \quad (6)$$

The above inequality is trivial when $\mu_i(x_i) = 0$. Hence (6) is equivalent to :

$$\bar{u}_i(\delta_{x_i}, \mu_{\hat{i}}) \geq \bar{u}_i(\delta_{y_i}, \mu_{\hat{i}}) \quad \text{all } i \in N, \text{ all } x_i \in [\mu_i], \text{all } y_i \in X_i \, (7)$$

Suppose now that μ is a mixed N E outcome of G. Then by Theorem 2 Chapter IV system (7) holds. Conversely (7) implies that $\bar{u}_i(\delta_{x_i}, \mu_{\hat{i}})$ does not depend on $x_i \in [\mu_i]$ and is therefore equal to $\bar{u}_i(\mu)$. This concludes the proof of the second statement of Lemma 1.

Next from Nash's theorem (Theorem 1 Chapter IV) the set NE (G_m) is non empty, which implies the non-emptyness of C E (G_m). Convexity and compactness of the latter set follow from our above remark that CE(G_m) is defined by a family of linear closed inequalities.

■

The argument justifying that a correlated equilibrium outcome is a self-enforcing agreement is identical to that supporting the cooperative view of the Nash equilibrium outcomes : actually by Lemma 1 a N E outcome, whether of the initial game or of its mixed version, is identified with a correlated equilibrium L where the probability distribution L is that of a N-uple of independent random individual strategies. In that case really no correlation of the strategies of the various players takes place.

One first, easy, way to achieve some positive correlation of individual strategies, is to take a convex combination of N E outcomes. Typically a lottery L such as :

$$L = \sum_{\alpha=1}^{p} \lambda_\alpha \delta_{x_\alpha}, \quad \sum_{\alpha=1}^{p} \lambda_\alpha = 1, \quad \lambda_\alpha \geqslant 0 \quad \text{all } \alpha \qquad (8)$$

where δ_x is the lottery with weight 1 on outcome x and x_α is a (pure) N E outcome of G, is a correlated equilibrium of G (notice that convex combinations of mixed N E outcomes also are in $CE(G_m)$). The red lights signal (Example 3 bis) is such a CE outcome. In many games, however, the set of correlated equilibrium outcomes is far broader than the mere convex hull of Nash equilibria   This crucial observation is illustrated by our next example.

Example 4. Musical chairs.

In this non standard version of Musical chairs, we have two players and three chairs marked 1, 2, 3. Each player' strategy is to pick a chair. Every body looses if both players pick the same chair. If, on the contrary, the two choices differ, this player - say i - whose chair immediately follows that of j wins twice more than j (with the convention that 1 follows 3). Hence the bimatrix game:

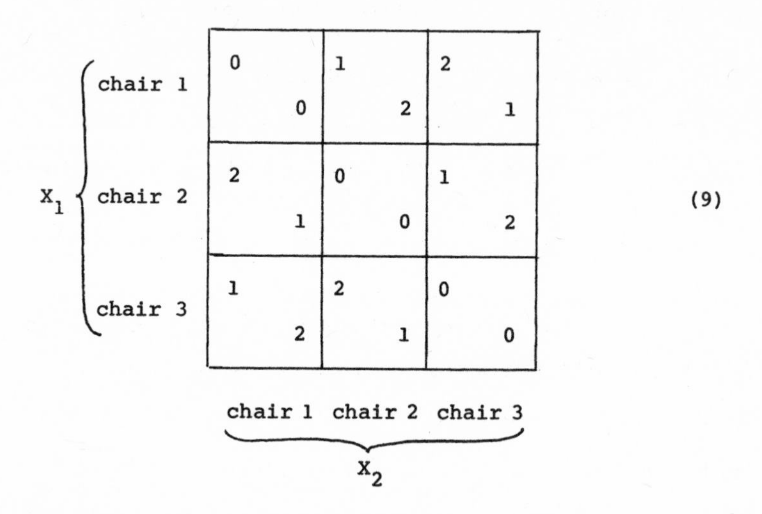

X_1 \ X_2	chair 1	chair 2	chair 3
chair 1	0 / 0	1 / 2	2 / 1
chair 2	2 / 1	0 / 0	1 / 2
chair 3	1 / 2	2 / 1	0 / 0

(9)

The initial game has no Nash equilibrium outcome. There is a unique, completely mixed, Nash equilibrium namely :

$$\mu_1^* = \mu_2^* = \frac{1}{3}\,\delta_1 + \frac{1}{3}\,\delta_2 + \frac{1}{3}\,\delta_3$$

This symmetrical outcome yields pay-off 1 to both players and is Pareto dominated. This is so because in (μ_1^*, μ_2^*) a bad deterministic outcome (i,i) is drawn with probability $\frac{1}{3}$. Consider now the following lottery L over $X_1 \times X_2$:

$$L(x_1, x_2) = \frac{1}{6} \quad \text{if} \quad x_1 \neq x_2$$
$$\qquad\quad = 0 \quad \text{if} \quad x_1 = x_2$$

L =

0	$\frac{1}{6}$	$\frac{1}{6}$
$\frac{1}{6}$	0	$\frac{1}{6}$
$\frac{1}{6}$	$\frac{1}{6}$	0

We claim that L is a correlated equilibrium of game (9). Suppose for instance that the deterministic outcome (2,3) has been selected and player 1 is therefore told to use strategy 2. Given L, player 1 infers from this message that player 2 is told to use either one of the strategies 1 or 3, each one of these with probability $\frac{1}{2}$. In other words assuming that player 2 obeys the signalled strategy amounts for player 1 to assume that player 2 uses the mixed strategy $\mu_2 = \frac{1}{2}\delta_1 + \frac{1}{2}\delta_3$. To the latter his best reply actually is to use strategy 2 since :

$$\bar{u}_1(\delta_2, \mu_2) = \frac{3}{2} > \bar{u}_1(\delta_1, \mu_2) = 1 > \bar{u}_1(\delta_3, \mu_2) = \frac{1}{2}$$

Similarly player 2 infers from the current signal (play strategy 3) that the signal to player 1 is strategy 1 with probability $\frac{1}{2}$ or strategy 2 with probability $\frac{1}{2}$. Next his best reply to player 1' mixed strategy $\mu_1 = \frac{1}{2}\delta_1 + \frac{1}{2}\delta_2$ happens to be strategy 3.

In view of the symmetry of our game, we can derive the self-enforcing property (3) of L for any possible drawing of the lottery.

Notice that L yields the Pareto optimal and fair pay-offs $(\frac{3}{2}, \frac{3}{2})$, which gives a justified incentive to co-operation by means of correlation.

The crucial feature of the non-binding agreement to obey the signal emitted by L bears on the distribution of information : both players are aware of the lottery L, upon which they have cooperatively agreed ; however after outcome x has been drawn by the lottery, player i is informed <u>only</u> of the i-th component x_i ; he can not observe the actual signal x_j received by any other player j : he must infer from L that the probability distribution of $x_{\hat{i}}$ is L_{x_i}.

Of course if L takes the form (8) with $(x_i)_\alpha \neq (x_i)_{\alpha'}$ all i, all $\alpha \neq \alpha'$, then $L_{x_{i,\alpha}}$ is indeed concentrated on x_α so that the mere signal $x_{i,\alpha}$ is enough for player i to know exactly the current signal to the other players. In that case, the self-enforcing property (3) requires that each x_α is itself a N E outcome of the initial game (see Example 3 bis). Our musical chair example makes clear that a skilful occultation of certain information channels can prove cooperatively profitable.

Remark 1.

As a natural generalization of Definition 2, we could allow any coalition T to jointly deviate by building up a correlated lottery L'_T over X_T and extend the strong equilibrium property in this random framework. This would lead to the notion of a "strong correlated equilibrium" for which however no general existence result can be expected : see for instance the two-person game of Example 5 below where no correlated equilibrium is Pareto optimal as well.

In our next example correlated equilibria do not allow any improvement of the, unique, Nash equilibrium. Nonetheless a more binding form of self-enforcing agreement relying on correlation again, is the accurate cooperative device.

Example 5. Competition by differentiation.

Two duopolists compete by choosing the quality of the good they supply on the same market. The game is symmetrical and each firm can choose among three strategies : Low, Medium or High quality. If both supply a low quality product or both supply a high one they both receive a zero profit. If one player offers medium quality whereas the other is at high or low, the medium player gets a profit of 2 whereas if both goods are medium, each player receives 1. To achieve the maximal joint profit the two firms must offer one high quality good and one low quality, in which case the high player receives 3 whereas the low player receives 1 . Notice that this model could be justified as an Hoteling location game.

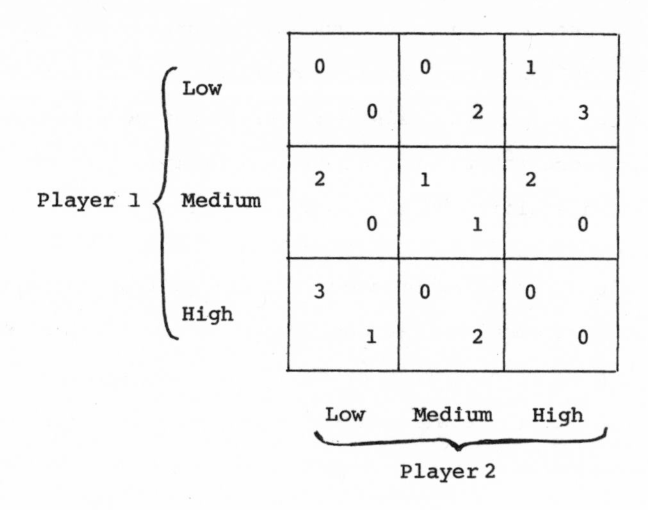

For non-cooperative players, this game is dominance-solvable : strategies Low are dropped first, next strategies High. This implies that (M , M) is the unique N E outcome of the initial game as well as its mixed extension .

Also the unique correlated equilibrium outcome is the lottery with mass 1 on (M , M) (to check this claim, remark that a CE lottery gives weight zero to any strictly dominated strategy, then apply this remark twice).

However the Pareto optimal pay-off vector (2, 2) can be achieved by the following agreement : let us construct a lottery with probability distribution :

$$L = \frac{1}{2}\,\delta_{(H,L)} + \frac{1}{2}\,\delta_{(L,H)}$$

Then each player, secretly and independently, sends a <u>binding</u> signal s_i to a neutral referee hired by both players : four signals are available to each player : the three pure strategies and signal OB (obedience to the lottery). After a pair (s_1, s_2) of messages are sent, the referee draws at random an outcome (x_1, x_2) according to lottery L. The final outcome is given by the following rule (enforced by the referee) :

$$\begin{cases} (x_1, x_2) & \text{if} \quad s_1 = s_2 = OB \\ (x_1, s_2) & \text{if} \quad s_1 = OB,\ s_2 = L, M, H \\ (s_1, x_2) & \text{if} \quad s_1 = L, M, H,\ s_2 = OB \\ (s_1, s_2) & \text{if} \quad s_i = L, M, H \quad i = 1, 2 \end{cases}$$

In words signal OB is a commitment to play that strategy selected by the lottery, taken by a player <u>before</u> the lottery is drawn. Next a signal like M is just the usual pure strategy M. As for correlated equilibria the pair of obedient signals is a self-enforcing agreement, namely :

$$\begin{cases} \bar{u}_1(OB,OB) = \frac{1}{2}u_1(H,L) + \frac{1}{2}u_1(L,H) = 2 > \bar{u}_1(y_1,OB) = \frac{1}{2}u_1(y_1,L) + \frac{1}{2}u_1(y_1,H) \\ \qquad\qquad\qquad \text{for all } y_1 \text{ in } \{L, M, H\} \\ \bar{u}_2(OB,OB) = \frac{1}{2}u_2(L,H) + \frac{1}{2}u_2(H,L) = 2 > \bar{u}_2(OB,y_2) = \frac{1}{2}u_2(H,y_2) + \frac{1}{2}u_2(L,y_2) \\ \qquad\qquad\qquad \text{for all } y_2 \text{ in } \{L, M, H\} \end{cases}$$

Contrary to correlated equilibria, the decision to obey the lottery can not be revised after a particular outcome has been drawn : it must be made once and for all before the random drawing.

<u>Definition 3</u>. (Moulin - Vial [1978]).

Notations and assumptions as in Definition 2. For all $i \in N$, we denote by $L_{\hat{i}}$ the marginal distribution of L over $X_{N\setminus\{i\}}$ namely :

$$L_{\hat{i}}(x_{\hat{i}}) = \sum_{x_i \in X_i} L(x_i, x_{\hat{i}}) \quad \text{all } x_{\hat{i}}$$

We say that lottery L is a <u>weak correlated equilibrium of</u> G if the following inequalities hold :

$$\forall i \in N \quad \forall y_i \in X_i \quad \sum_{x \in X_N} u_i(x) \; L(x) \geq \sum_{x_{\hat{i}} \in X_{N\setminus\{i\}}} u_i(y_i, x_{\hat{i}}) \; L_{\hat{i}}(x_{\hat{i}})$$

We denote by $WCE(G)$ the set of weak correlated equilibriums of G.

Lemma 2.

The set $WCE(G)$ is a non-empty convex compact subset of the unit simplex of $\mathbb{R}^{X_N}$. It contains the set $CE(G)$ of correlated equilibria of G as a subset.

If μ is an outcome of the mixed game G_m, then the associated lottery $L = \bigotimes_{i\in N} \mu_i$ is a weak correlated equilibrium of G if and only if μ is a Nash equilibrium of G_m.

Proof.

By Definition 2, lottery L belongs to $WCE(G)$ if and only if :

$$\forall i\in N \quad \forall y_i \in X_i \quad \sum_{x\in X_N} u_i(x)L(x) \geq \sum_{x\in X_N} u_i(y_i, x_{\hat{i}})\, L(x)$$

These inequalities are obtained from system (4) by keeping i and y_i fixed while summing up over $x_i \in X_i$. Hence the inclusion :

$$CE(G) \subset WCE(G)$$

The other statements of Lemma 2 are obvious.

■

Definition 2 represents the last step in the conceptual development of self-enforcing agreements using randomization and mutual secrecy. Starting with the Nash equilibrium in pure strategies, we allow first for independent randomizations of individual strategies - mixed strategies - and observe that if the players can draw these private lotteries in mutual secrecy, self-enforcing agreements always exist (Theorem 1 Chapter IV). Next if the players can build correlated lotteries sending decentralized signals to individuals the set of equilibria becomes convex. From Lemma 1 and 2 we deduce the inclusions :

$$NE(G) \subset NE(G_m) \subset CE(G) \subset WCE(G)$$

From left to right more and more informational constraints must be met to make an equilibrium outcome self-enforcing. For a NE or a mixed NE outcome, only the privacy of individual strategies has to be preserved. For a CE outcome, we must, in addition, prevent individual players from observing anything but their own component of the actual drawing of the correlated lottery. For a WCE outcome, a neutral referee must draw an outcome of which nothing is revealed to any player, next ask independently and secretly to each player whether or not he is willing to play this "blind" outcome, inform only those players who voluntarily choose to obey the lottery of the actual drawing of the lottery, and force the declared obedient players to faithfully play the commanded strategy. In short the referee must have the power to block any strategic use of the actual drawing of the lottery.

The common feature of these increasingly demanding informational scenarios is that after an agreement has been reached on the choice of a particular correlated strategy, no more direct communications can take place among players. In case of repeated drawing of the agreed upon lottery, cooperation becomes implicit and is hardly recognisable. This form of tacit collusion is examplified in the literature on oligopoly behaviour : "Among all possible arrangements which can be designed in an oligopoly, let us single out those which avoid explicit communication between the participants. One example of such arrangement is present in the "electrical equipment conspiracy of the 1950 s" (Sherer [1970]) which involved 29 U.S. companies selling different sorts of products. For the power-switching equipment, sealed-bid competitions were sponsored by the U.S. government. Through some preliminary secret negotiation, each other was assigned a share of all sealed-bid business. The sellers then coordinated their bidding so that each one of them was low bidder in just enough transactions to gain its predetermined share of the market. This was achieved by dividing the territory into four quadrants assigned different sellers to each quadrant and letting the sellers in a quadrant rotate their bids. A "phases of the moon" system was used to decide of the low-bidding privileges. Thus, the result was an ostensibly random process conveying the impression of independent behavior." (Gerard-Varet and Moulin [1978])

Exercise 4. In the musical chair game (Example 4) prove that all four sets $NE(G)$, $NE(G_m)$, $CE(G)$ and $WCE(G)$ are different. Prove that the Pareto optimal CE pay-offs cover the interval $[(\frac{5}{3}, \frac{4}{3}), (\frac{4}{3}, \frac{5}{3})]$ whereas the Pareto optimal WCE pay-offs cover $[(2,1), (1,2)]$.

Exercise 5. Let $G = (X_i, u_i; i \in N)$ be such that X_i has two elements for all $i \in N$. Prove that $CE(G) = WCE(G)$.

Problem 1. Weak correlated and Nash equilibrium.
Moulin [1976].

We fix a two person game $G = (X_1, X_2, u_1, u_2)$ with finite strategy sets X_1, X_2 and denote by $G_m = (M_1, M_2, \bar{u}_1, \bar{u}_2)$ the associated mixed game. Given a correlated lottery L, we denote by L_1, L_2 (instead of $L_{\hat{2}}, L_{\hat{1}}$) the marginal distributions of L respectively over X_1, X_2. Remark that $L_i \in M_i$, $i = 1, 2$. Finally we denote by $[u_i, L]$ the expected utility from lottery L to player i, $i = 1, 2$.

1) An outcome (μ_1^*, μ_2^*) of G_m is said to be a Nash-plus equilibrium if there exists no lottery L such that :

$$\begin{cases} \bar{u}_i(\mu_i^*, L_j) \leq [u_i, L] & \text{for } i = 1, 2, \quad j \neq i \\ \text{with at least one strict inequality.} \end{cases}$$

Interpretation : for all conceivable agreements L either one player has an incentive to stick to his N-plus E strategy ($\bar{u}_i(\mu_i^*, L_j) > [u_i, L]$) or both players are indifferent between their N-plus E strategies and being faithful to the agreement, given that the other himself is faithful ($\bar{u}_i(\mu_i, L_j) = [u_i, L]$ all $i = 1, 2, \quad j \neq i$).

Prove that a N-plus E outcome is a mixed N E outcome.

2) We denote by $N_+E(G_m)$ the set of N-plus E outcomes of G_m, and by N_+E_i the projection of this set over M_i. Prove that :

$$N_+E(G_m) \subset N_+E_1 \times N_+E_2 \subset NE(G_m) \tag{10}$$

Hint : if (μ_1, μ_2) and (ν_1, ν_2) are in $N_+E(G_m)$ prove that :

$$\bar{u}_1(\mu_1, \nu_2) = \bar{u}_1(\mu_2, \nu_2)$$

$$\bar{u}_2(\mu_1, \nu_2) = \bar{u}_2(\mu_1, \nu_1)$$

3) Prove that an equilibrium in dominating strategy is a N-plus E outcome as well :

$$D_1(u_1) \times D_2(u_2) \subset N_+E(G_m)$$

Prove that if $N_+E(G_m)$ is non-empty, then

$N_+ E_1 \times N_+ E_2$ possesses a Pareto dominating element μ^* :

$$\begin{cases} \forall \mu \in N_+ E_1 \times N_+ E_2 : \forall i = 1,2 \ \ \bar{u}_i(\mu) \leq \bar{u}_i(\mu^*) \\ \mu^* \in N_+ E_1 \times N_+ E_2 \end{cases}$$

4) Prove that $\mu^* = (\mu_1^*, \mu_2^*)$ is a Nash - plus equilibrium outcome of G_m if and only if there exists a real number λ, $0 < \lambda < 1$, such that :

$$\forall (x_1,x_2) \in X_1 x X_2 : \lambda u_1(x_1,x_2) + (1-\lambda) u_2(x_1,x_2) \leq \lambda \bar{u}_1(\mu_1^*,x_2) + (1-\lambda)\bar{u}_2(x_1,\mu_2^*) \quad (11)$$

(where by a slight abuse in notations, $\bar{u}_i(\mu_i, x_j)$ stands for $\bar{u}_i(\mu_i, \delta_{x_j})$).

Deduce that $N_+ E(G_m)$ as well as $N_+ E_i$, $i = 1, 2$ are compact.

5) From now on we assume that u_i depends really on x_i for $i = 1, 2$.

$$\exists \ x_i, x_i' \in X_i \quad x_j \in X_j \quad u_i(x_i, x_j) \neq u_i(x_i', x_j) \qquad (12)$$

We denote by $W^\circ C E(G_m)$ the (possible empty) interior of $WC E(G_m)$ relative to the unit simplex of $\mathbb{R}^{X_1 x X_2}$. Prove the following equivalence :

$$L \in W^\circ C E(G_m') \Leftrightarrow [\bar{u}_i(x_i, L_j) < [u_i, L] \text{ all } i = 1,2, j \neq i, \text{ all } x_i \in X_i \quad (13)$$

6) Suppose, in addition to (10), that no player has a dominating strategy in G_m. Then prove the following equivalence :

$$W° \cap E \neq \phi \Leftrightarrow N_+ E = \phi$$

<u>Hint</u> : Setting $Z = X_1 \cup X_2$, the disjoint union of X_1 and X_2, consider the real valued function f defined over $X_{\{12\}} \times Z$ as follows :

$$\forall (x_1, x_2) \in X_{\{12\}} \ \forall z \in Z \ f((x_1,x_2),z) = u_1(x_1,x_2) - u_1(z,x_2) \text{ if } z \in X_1$$

$$= u_2(x_1,x_2) - u_2(x_1,z) \text{ if } z \in X_2$$

Then observe that the mixed value ν of the game $(X_{\{12\}}, Z, f, -f)$ is non-negative. Show that ν is zero if and only if for some λ, $0 < \lambda < 1$, inequalities (11) hold. Next show that ν is positive if and only if there exists a lottery L such that the right hand inequalities in (13) hold.

REFERENCES

AUMANN, R.J. 1959. "Acceptable points in general cooperative n-person games in Contributions to the theory of games",IV, Annals of Maths. Studies 40. Princeton University Press.

AUMANN, R.J. 1974. "Subjectivity and correlation in randomized strategies". Journal of Mathematical Economics 1 : 67-96.

GERARD-VARET, L.A. and H. MOULIN. 1978. "Correlation and duopoly", Journal of Economic Theory 19, 1 : 123-149.

LUCE, R.D. and H. RAIFFA. 1957. Games and decisions. New York: J. Wiley and Sons.

MOULIN, H. 1976. "Correlated and prudent equilibria of a two-person game". Cahiers de Mathématiques de la Décision n° 7605 . Paris : Université Dauphine.

MOULIN, H. and J.P. VIAL. 1978. "Strategically zero-sum games". International Journal of Game Theory 7, 3/4 : 201-221.

SHERER, F.M. 1970. Industrial pricing. Chicago : Rand Mac Nally.

CHAPTER VI. STABILITY BY THREATS.

Our players, willing to cooperate and aware of the strategic interdependency described by the normal form of the game, anticipate now the reactions by the others to their own strategical moves. By threatening each other they can actually stabilize a considerable subset of outcomes ; thus we view a threat as a cooperative device.

Cooperation by mutually deterring threats is an old topic in the oligopoly literature. For instance in a Cournot duopoly model (such as Example 5 Chapter III) the maximal joint profit outcome is not a Nash equilibrium since by unilaterally increasing his own supply a firm improves upon its instantaneous profit. However such a move presumably induces the other firm to a similar improvement that eventually leaves both players worse off than at the initial outcome. As Sherer [1970] points out "each would be reluctant to take measures, which when countered, would leave all members of the industry worse off". (see also Example 5 below).

Deterrence is a powerful cooperative tool. To enforce the stability of an agreed upon outcome, our players threaten each other, i.e. announce a specific reaction to potential deviations : if the deviant player happens to be worse off after the announced threat has been carried out, he is deterred from entering the deviation, and the non-binding agreement proves to be stable.

Thus deterrence is the "skilfull exploitation of potential forces": a successful threat is one that is not carried out. (Schelling [1971])

Whereas the self-enforcing agreements of Chapter V require strict privacy of the final strategic decision, on the contrary a threat is effective only if any deviation is publicly known. Henceforth stability by deterrence demands transparence of the individual strategic choices. Military and economic examples of the limitation of strategic privacy for the sake of cooperative stability abound (see Section 1).

The main equilibrium concept here is the α- core, i.e. the set of those outcomes that can be immunized against all coalitional deviations by appropriate threats (Section 2). If only individual deviations are feasible, deterring threats stabilize any imputation of the game (Section 1).

When the α - core is empty, more complex deterring scenarios, involving "counter threats", are necessary to stabilize some imputations against coalitional deviations (Section 2). On the contrary, if the α- core is non empty, several smaller subsets of it are derived by imposing more conditions on the deterring threats : one such subset is the β - core, that is deeply connected with the analysis of repeated games (Section 3). Next in Section 4 we define another subset of the α - core namely the s - core of a two person game. The s - core is made up of those imputations stabilized by credible threats, i.e. best reply behaviour. This leads to a qualitative classification of two-person games.

In Section 5 the alternative formalism of games in characteristic form (with side-payments) is presented : there the core stability set is defined by finitely many linear constraints and leads to simple computational results.

I - IMPUTATIONS

Example 1. Cooperative Solution of the Prisonners' Dilemma.

The non cooperative outcome of Prisonners' Dilemma (Example 1 Chapter I) is war : (A_1, A_2). To cooperatively enforce the peaceful outcome (P_1, P_2), both players announce the following tit-for tat behaviour :

$$\begin{cases} \text{if you play peaceful, I shall play peaceful too} \\ \text{if you play agressive, I shall play agressive too} \end{cases} \quad (1)$$

Given that my opponent threatens me in this way, I shall better be peaceful (since this leads us to peace (P_1, P_2)) rather than agressive (which leads to war (A_1, A_2). This is the stability property of threat (1).

Now an odd feature of bilateral threats (1) emerges: to implement behaviour (1) a player must act as a follower, so that if each player chooses his strategy once for all, at most one of them can possibly be in the position of a follower. Moreover the prisonners' dilemma game is symmetrical and we want the cooperative outcome (peace) to be enforced by a deterring scenario where our players have symmetrical roles.

To overcome this difficulty we can imagine that the game is played repeatedly and the overall pay-off results from a suitable summation of the instantaneous pay-offs ; presumably if the short-run profit of a non-cooperative deviation is overweighted by the long-run losses implied by reaction (1), then the deterring scenario succeeds in enforcing the peaceful cooperative outcome (see Section 3 below). Here we formalize cooperation by threats by means of a more simple mathematical object, consisting only of an agreed upon-outcome and a threat to deter each individual player to betray this outcome.

Definition 1.

Let $G = (X_i, u_i ; i \in N)$ be a given N - normal form game. A deterring scenario is a $(N+1)$-uple $(x, \xi_{\hat{i}} ; i \in N)$ where $x \in X_N$ is an outcome of G and for all $i \in N$, $\xi_{\hat{i}}$ is a threat against player i, that is a mapping from X_i into $X_{N \setminus \{i\}}$ such that :

$$\begin{cases} \xi_{\hat{i}}(x_i) = x_{\hat{i}} \\ \forall y_i \in X_i \setminus \{x_i\} : u_i(y_i, \xi_{\hat{i}}(y_i)) \leq u_i(x) \end{cases} \tag{2}$$

We interpret a deterring scenario in the framework of a game played over an infinite time period. At one particular date each player plays a particular strategy ; he can change at will his strategy at any time. The strategies of all players are constantly public. This is the crucial informational feature that prevents any player from deviating secretly.

A player faithful to the agreement plays first his agreed upon strategy x_i and observes the current strategies $y_{\hat{i}}$ used by other players. As long as $y_{\hat{i}} = x_{\hat{i}}$, player i keeps strategy x_i ; as soon as some individual player, say j, switches to strategy $y_j \neq x_j$, then i plays the i-th component of $\xi_{\hat{j}}(x_j)$ <u>for ever</u>. The stability condition (2) means that, if all other players are faithful to the deterring scenario, no individual player has any incentive to be (alone) unfaithful. Namely it is understood that the pay-off over an infinite period of time always overweights that over any finite length period.

Definition 1 considers only deviation by single players : it will be generalized in the next section so as to allow for coalitional deviations (see Definition 3).

Surely the informational constraint that all strategic moves must occur publicly might be difficult to achieve. For instance the modern armament technology makes surprise attack more and more frightening. Thus a demilitarized no man's land on which all agressive moves are visible, or a nuclear sites inspection agreement are both informational devices pertaining to deterring threats. As another example, an "open price agency" prevents competing firms from offering secret discounts to their customers and therefore sweetens the overall competition (Sherer [1970]). On the other hand the anglo-japanese agreement to limit the production of war-crusers did not stipulate any control clause as it was at the time judged impossible to build up such objects secretly (see Aron [1962]).

Lemma 1.

Notations as in Definition 1.

Let $(x, \xi_{\hat{i}}\, ; i \in N)$ be a deterring scenario. Then x is an individually rational outcome :

$$\sup_{y_i} \inf_{y_{\hat{i}}} u_i(y_i, y_{\hat{i}}) \leqslant u_i(x) \qquad \text{all } i \in N \tag{3}$$

Conversely suppose X_i is compact and u_i continuous, all i. Then G possesses at least one individually rational outcome. For any such outcome x there exists for all $i \in N$ a threat $\xi_{\hat{i}}$ against i such that $(x, \xi_{\hat{i}}\, ; i \in N)$ is a deterring scenario.

Proof.

From (2) we deduce :

$$\inf_{y_{\hat{i}}} u_i(y_i, y_{\hat{i}}) \leqslant u_i(y_i, \xi_{\hat{i}}(y_i)) \leqslant u_i(x)$$

for all $y_i \in X_i$. Hence the first statement of Lemma 1. Conversely we know that under our topological assumption each player has at least one prudent strategy say x_i (see Lemma 1 Chapter I) ; then outcome $x = (x_i)_{i \in N}$ is individually rational.

Next for any i and $y_i \in X_i$, $y_i \neq x_i$, we pick an element $y_{\hat{i}} = \xi_{\hat{i}}(y_i) \in X_{N \setminus \{i\}}$ such that :

$$u_i(y_i, y_{\hat{i}}) = \inf_{z_{\hat{i}}} u_i(y_i, z_{\hat{i}}) \leqslant \sup_{z_i} \inf_{z_{\hat{i}}} u_i(z_i, z_{\hat{i}}) \leqslant u_i(x)$$

This concludes the proof of Lemma 1. ■

Definition 2.

An imputation of game $G = (X_i, u_i; i \; N)$ is a Pareto optimal individually rational outcome. We denote by $I(G)$ the set of imputations of G.

Lemma 2.

If X_i are compact and u_i continuous for all $i \in N$, the game G possesses at least one imputation.

Proof.

Parallels that of Lemma 3 Chapter I.

Denote by $IR(G)$ the non-empty compact subset of individually rational outcomes of G. Next choose an element x of $IR(G)$ that maximizes the function $\sum_{i\in N} u_i$ over $IR(G)$. Then x is a Pareto optimal outcome. Suppose on the contrary that y Pareto dominates x. Then y belongs to $IR(G)$ and is such that $(\sum_{i\in N} u_i)(x) < (\sum_{i\in N} u_i)(y)$, a contradiction.

∎

An imputation is together a Pareto optimum and gives to each player at least his secure utility level. By Lemma 1 all imputations result from a Pareto optimal deterring scenario, that is to say a non-binding agreement stable with respect to individual deviations as well as objections by the grand coalition N (Pareto optimality). Conversely two minimal requirements for cooperative agreements are individual rationality (since no player can be forced below his secure utility level as long as he keeps control of his own strategy) and Pareto optimality (since agreements are reached by the unanimous consent of all players). Thus I(G) is the maximal range for cooperative negotiation. If G is inessential (or almost inessential : see Problem 1, Chapter III), the set I(G) is a singleton and we are home. But in most games I(G) encompasses several outcomes among which the conflict can be wild.

Example 2. A trading game.

Player 1 sells an indivisible object to player 2 and must decide to charge a high price or a low one. At both prices player 2 - the buyer - still makes a profit : he can not bargain the price, only refuse the deal.

Player 1 \ Player 2	Buy	Refuses
High	2, 1	0, 0
Low	1, 2	0, 0

Outcome (High, Buy) is the equilibrium in dominating strategy ; it is also a Pareto optimum. Although the unambiguous output if the players are not able to communicate, this outcome is not the only imputation : (Low, Buy) is the other one. To win (i.e. to sell the object at high price), player 1 tries to commit himself to sell at a high price, i.e. acts as a leader. On the other hand player 2 wins by convincingly threatening the seller : I shall by at a low price but will refuse the deal at a high price. This seemingly irrational behaviour (since the buyer refuses a still profitable deal), proves very rational if the threat is believed by the seller. From a cooperative point of view we do not discriminate between two deterring scenarios that is to say we do not arbitrate the conflict between these two antagonistic threats (commitment to High by the seller, refuses to buy unless at a low price by the buyer) that can not be simultaneously spelled out except at the risk to end up at a Pareto dominated outcome (no trade).

In the above example, the buyer wins by enhancing the "radical" threat : to any deviation of yours, I will react by minimizing your pay-off (i.e. I will play the worst reply to your eye). This radicalism may imply that carrying out the threat is harmful to both players : announce by the buyer that he will refuse to deal at a high price is, he hopes, a purely deterring signal but to make it credible it must be likely that this threat will actually be carried out ; in this sense even the most convincing and successful threat is risky if the announced reaction to deviations do not always coïncide with a best reply of the threatening player. As an illustration, compare the "agressive" threat and the "warning" in Lemma 3 below. The issue of credibility and genuine risk of the deterring threats is analyzed for two person - games in Section 4 below.

The situation of Example 2 is easily generalized to all two person - games.

We consider (X_1, X_2, u_1, u_2), a two person game where X_i are compact and u_i continuous, $i = 1, 2$. A favourite imputation of player 2 is an outcome x^i such that :

$$x^i \in I(G) \quad u_i(x^i) = \sup_{x \in I(G)} u_i(x)$$

which is equivalently written as :

$$u_i(x^i) = \sup \{u_i(x) / u_j(x) \geq \sup_{y_j} \inf_{y_i} u_j(y_j, y_i)\} \quad (4)$$

Moreover any two solutions to (4) yield the same pay-off to both players :

$$[\ x \text{ satisfies } (4)] \Rightarrow [\ u_j(x) = u_j(x^i) \quad j = 1,2\ ]$$

The proof of these two assertions is left as an exercize to the reader.

Lemma 3.

Let x^i be an imputation such that (4) holds.

Let ξ_i be an (agressive) threat of player i :

$$\begin{cases} \xi_i(x^i_j) = x^i_i \\ \forall y_j \in X_j \setminus \{x^i_j\} \quad u_j(y_j, \xi_i(y_j)) = \inf_{y_i} u_j(y_j, y_i) \end{cases} \tag{5}$$

Let ξ_j be a warning of player j , i.e. a threat such that :

$$\begin{cases} \xi_j(x^i_i) = x^i_j \\ \forall y_i \in X_i \setminus \{x^i_i\} \quad u_j(\xi_j(y_i), y_i) = \sup_{y_j} u_j(y_j, y_i) \end{cases} \tag{6}$$

Then (x^i, ξ_i, ξ_j) is a deterring scenario.

Proof.

From (5) we deduce

$$u_j(y_j, \xi_i(y_j)) \leq \sup_{z_j} \inf_{z_i} u_j(z_j, z_i) \text{ all } y_j \in X_j \setminus \{x^i_j\}$$

Since x^i is individually rational (in short i.r.), this implies :

$$u_j(y_j, \xi_i(y_j)) \leq u_i(x) \quad \text{all } y_j \in X_j \setminus \{x^i_j\}$$

On the other hand (6) yields :

$$u_j(\xi_j(y_i), y_i) \geq \inf_{z_i} \sup_{z_j} u_j(z_j, z_i), \text{ all } y_i \in X_i \setminus \{x^i_i\}$$

Fix now a strategy y_j, $y_j \neq x_j^i$, and suppose :

$$u_i(\xi_j(y_i), y_i) > u_i(x^i)$$

The latter two inequalities and the fact that x^i is i.r., imply that $y = (\xi_j(y_i), y_i)$ itself is an i.r. outcome. By our topological assumption there exists a Pareto optimal outcome z above y :

$$u_i(y) \leqslant u_i(z) \quad , \quad u_j(y) \leqslant u_j(z)$$

Thus z is an imputation such that $u_i(x^i) < u_i(z)$, a contradiction.

We have proved finally :

$$u_i(\xi_j(y_i), y_i) \leqslant u_i(x^i) \quad \text{all } y_i \in X_i \setminus \{x_i^i\}$$

■

As suggested by Schelling [1971] we call warning a threat which coïncides with the induced behaviour of a follower, a very convincing reply indeed. Any other threat has something of a "doomsday machine" : at the time of carrying out the threat, no rational short-sighted player would do it, hence its success as a deterring device requires binding commitment by the threatening player or at least belief of this commitment by the threatened player.

Lemma 3 emphasizes that the concept of deterring scenario does not allow any discrimination among the imputations of a particular game. It says that in a two person game, if one lion-player has the power to spell out convincingly any threat, whereas his sheep-partner is not believed to have the nerves to ever play anything but a best reply strategy, then the lion keeps the whole surplus of cooperation for himself.

To circumscribe a smaller subset of $I(G)$ we take next into account coalitional threats.

Exercise 1. Metagames. (Howard [1971])

Given a two-person game $G = (X_1, X_2, u_1, u_2)$ with finite strategy sets we denote by $S(1;G)$ its extension where player 1 acts as a leader (i.e. the game denoted $\tilde{G}$ in Lemma 5, Chapter II Section 4) :

$$S(1;G) = (X_1^{X_2}, X_2; \tilde{u}_1, \tilde{u}_2)$$

Next consider the game :

$$H = S(2; S(1, G))$$

1) Interpret this game.

2) Prove that a pair (a_1, a_2) is a NE pay-off vector of H if and only if the following two properties hold :

i) (a_1, a_2) is a feasible payoff vector of G : for some $x^\star \in X_{\{12\}}$: $(a_1, a_2) = (u_1(x^\star), u_2(x^\star))$

ii) $\inf_{x_2} \sup_{x_1} u_1 \leq u_1(x^\star)$

$\sup_{x_2} \inf_{x_1} u_2 \leq u_2(x^\star)$

3) Prove that the 1-Stackelberg payoff of $S(1 ; G)$ (see Lemma 6 Section IV below) is player 1'best imputation payoff in G. Compute also the i-Stackelberg payoff of H for i=1,2.

II. THE α - CORE.

Definition 3.

Given a N- normal form game $G = (X_i, u_i ; i \in N)$, the α-core of G is the subset, denoted $C_\alpha(G)$ of those outcomes $x^\star$ such that :

for all coalition T, $T \subset N$ and all joint strategy $x_T \in X_T$ there exists a joint strategy $x_{T^c} \in X_{T^c}$ such that :

$$\text{No} \begin{cases} u_i(x_T, x_{T^c}) \geq u_i(x^\star) & \text{for all } i \in T \\ u_i(x_T, x_{T^c}) > u_i(x^\star) & \text{for at least one } i \in T \end{cases}$$

For the sake of simplicity we have written Definition 3 without explicit reference to a deterring scenario where coalitions react to joint deviations by the complement coalition. Formally, such a coalitional deterring scenario would be :

$(x^{\star}, \xi_T, \text{all } T \subset N)$

where ξ_{T^c}, a mapping from X_T into X_{T^c}, is such that for no coalition $T \subseteq N$ and no joint strategy $x_T \in X_T$ we have :

$$\begin{cases} u_i(x_T, \xi_{T^c}(x_T)) \geq u_i(x^{\star}) & \text{for all } i \in T \\ u_i(x_T, \xi_{T^c}(x_T)) > u_i(x^{\star}) & \text{for some } i \in T \end{cases}$$

Then outcome $x^{\star}$ belongs to $C_{\alpha}(G)$ if and only if there exists for all $T \subseteq N$ a threat by coalition T^c against potential deviations by coalition T such that $(x^{\star}, \xi_T ; T \subseteq N)$ is a coalitional deterring scenario.

By definition 3 an outcome $x^{\star}$ is in the α-core of G if every joint deviation x_T by coalition T can be "countered" by a move x_{T^c} of the complement coaltion T^c that deters at least one member of T from entering deviation x_T since he is eventually worse off: $u_i(x_T, x_{T^c}) < u_i(x^{\star})$. (or else every member of T keeps the same utility level afterall :

$$u_i(x_T, x_{T^c}) = u_i(x^{\star}) \quad \text{all } i \in T \text{).}$$

Applying this property successively to $T = N$, and $T = \{i\}$, all $i \in N$, implies that an outcome in the α-core is in particular an imputation :

$$C_{\alpha}(G) \subset I(G)$$

Observe that a Pareto optimal NE outcome is in particular an imputation (that with the passive threat - no reaction - forms a deterring scenario) ; similarly a strong equilibrium is in the α-core as well (stabilized, again, by passive threats) :

$$NE(G) \cap PO(G) \subset I(G)$$

$$SE(G) \subset C_\alpha(G)$$

Exercise 3 below illustrates these inclusions. In particular a simple three players game is proposed where no strong equilibrium exists whereas the α-core is a doubleton (question 2).

In a two person game, the α-core coïncides with the set of imputations and therefore is never empty (under the mild topological assumptions of Lemma 2). It is a remarkable fact that in games with at least three players, the α-core can be the empty set.

Example 3. The Condorcet paradox.

Let N be a society with an odd number of players, who must pick one among a finite set A of candidates. The voting rule is plurality voting : each player casts a vote for one candidate and one among the candidates with the highest score wins.

For all $i \in N$ denote by u_i player i's utility over A . Indifferences are ruled out so that only p! different orderings can occur (where $p = |A|$ is the cardinality of A).

Agent i's strategy set is $X_i = A$ and the voting rule is a mapping π from X_N into A such that for all $x \in X_N$:

$$\pi(x) = a \Rightarrow |\{i \in N / x_i = a\}| \geqslant |\{i \in N / x_i = b\}| \text{ all } b \in A \quad (7)$$

Note that ties are broken arbitrarily. Given u_i, $i \in N$, our voters face the normal form game :

$$G = (X_i, u_i \circ \pi ; i \in N).$$

of which we seek the cooperatively stable outcomes. Suppose then that $x^* \in X_N$ belongs to the α-core of G, and denote $a = \pi(x^*)$. Remark that any coalition T with a strict majority of players, $|T| > \frac{N}{2}$, can force the election of any outcome b by simply casting a unanimous vote for b (this follows from (7) :

$$[x_i = b \text{ all } i \in T] \Rightarrow [\pi(x_T, x_{T^c}) = b, \text{ all } x_{T^c} \in X_{T^c}]$$

Therefore the property :

$$\forall i \in T \quad u_i(b) > u_i(a) \quad (8)$$

would violate the α-core stability property of x^* (see Definition 3). Thus (8) fails to be true for all strict majority T of players and all candidate $b, b \neq a$. This is equivalently written as :

$$|\{ i \in N / u_i(a) > u_i(b) \}| > \frac{N}{2} \quad \text{for all } b,\ b \neq a \qquad (9)$$

Namely since N is odd either a coalition is a strict majority or its complement is . Property (9) means that candidate a is a Condorcet winner of the preference profile u_i, $i \in N$, that is to say a candidate that defeats any other candidate in pairwise contests. We have just proved :

$$[x^{\star} \in C_{\alpha}(G)] \Rightarrow [\pi(x\ ) \text{ is a Condorcet winner of } (u_i)_{i \in N}]$$

The converse property is also true : it is left as an easy exercize to the reader.

To a given preference profile u_i, $i \in N$, corresponds at most one Condorcet winner (by (9)). Accordingly two cases only may arise :

<u>Case 1</u>. The profile $(u_i, i \in N)$ possesses a Condorcet winner a. Then the α-core of G is made up of all outcomes x such that $\pi(x) = a$:

$$C_{\alpha}(G) = \pi^{-1}(a)$$

<u>Case 2</u>. The profile $(u_i, i \in N)$ has no Condorcet winner. Then the α-core of G is empty.

$$C_{\alpha}(G) = \phi$$

Case 2 is known as the Condorcet paradox situation, and has been the subject of an extensive literature in the

Social Choice Theory (see e.g. Sen [1970] Moulin [1981]). To construct a preference profile with no Condorcet winner, assume $p \geq 3$ and $|N| = n \geq 5$. Next pick three candidates a, b, c and three integers n_1, n_2, n_3 such that:

$$\begin{cases} n_1 + n_2 + n_3 = n \\ n_k + n_\ell > n_m \quad \text{all } \{k,\ell,m\} = \{1,2,3\} \end{cases}$$

Then suppose society N splits into three homogeneous coalitions N_k, with cardinality n_k, $k = 1,2,3$ such that :

$$\begin{aligned} &\text{for all } i \in N_1 : u_i(a) > u_i(b) > u_i(c) > u_i(\alpha) \\ &\text{for all } i \in N_2 : u_i(b) > u_i(c) > u_i(a) > u_i(\alpha) \quad \text{all } \alpha \in A\setminus\{a,b,c\} \\ &\text{for all } i \in N_3 : u_i(c) > u_i(a) > u_i(b) > u_i(\alpha) \end{aligned}$$

Candidate a is preferred to b by the agents of $N_1 \cup N_3$. Since $n_1 + n_3 > n_2$, b is not a Condorcet winner. Similarly the strict majority $N_1 \cup N_2$ prefers b to c and the strict majority $N_2 \cup N_3$ prefers c to a.

Exercise 2.

Prove that notall outcomes x in the α- core of G are strong equilibria. However $SE(G)$ and $C_\alpha(G)$ are simultaneously empty or yield the election of the same candidate :

$$\pi(SE(G)) = \pi(C_\alpha(G))$$

How is Example 3 modified if $|N|$ is even ?

When the α-core of a game is empty, cooperative stability can not be achieved by deterring threats only. Even transparency of strategic deviations do not prevent coalitions from raising profitable deviations that can not be countered by an appropriate move of the non deviating players. To restore stability, we invoke a more subtle behavioural scenario where the reaction by the non-deviating players is to bribe some members of the betraying coalition in such a way that the non-bribed betrayers suffer eventually a loss.

Consider for instance the plurality voting game (Example 3) with three players and three candidates when the preferences involve a Condorcet cycle :

$$u_1(a) > u_1(b) > u_1(c)$$

$$u_2(b) > u_2(c) > u_2(a)$$

$$u_3(c) > u_3(a) > u_3(b)$$

Choose an outcome $x = (b, b, c)$ where b is the elected candidate. Stability of x is jeopardized by coalition $\{1, 3\}$: by jointly voting for a , player 1 and 3 enjoy a strictly better utility level, not threatened by any strategical reaction of player 2 :

$$u_i(a, x_2, a) > u_i(b, b, c) \text{ for } i = 1,3 \text{ and all } x_2 \in X_2$$

However, player 2 can offer still a better deal to player 3 : by both voting for c , player 2 and 3 force the election of a, thus guaranteing to player 3 his optimal utility level ; at the same time player 1' utility falls to his worst level. Anticipating that player 2 will bribe player 3, player 1 is deterred from entering the initially deviating coalition {1, 3}, since he ends up below his initial utility level : $u_1(c) < u_1(b)$.

The above two steps scenario, where a coalitional move generates a counter coalition bribing some players of the initial coalition to eventually deter the other deviating players, leads to a stability concept generalizing the α-core notion. Although the technical complexity of the threat and counter-threat scenario is high, and its descriptive power is debatable, the corresponding equilibrium set proves to be always non-empty (see Laffond-Moulin [1980]), a very desirable theoretical feature. In Problem 3 below we describe this result in the three players case.

Remark 1.

Several two step stability concepts are available in the game theory literature : the Von-Neumann - Morgenstern solution (Luce and Raïffa [1957] , Vickrey strong solution [1959] , the subsolutions and supercore Roth [1976]) ; for a survey of this literature, see Rosenthal [1972] .

Exercise 3.

1) A three players prisonners' dilemma

Each player can be agressive (A) or cooperative (C) and the game is symmetrical. The various payoffs are listed below :

(C, C, C) payoff vector (2, 2, 2)

(A, C, C) payoff vector (3, 1, 1)

(A, A, C) payoff vector (2, 2, 0)

(A, A, A) payoff vector (1, 1, 1)

For instance the three players are three firms competing by setting a regular price (C) or using a dumping policy (A). The maximal joint profit is 6 (when all players are cooperative) and decreasesby one unit per agressive player: each player switching from C to A gains one additional unit of profit and decreases by one the profit of the other two.

Prove that the dominating strategy equilibrium is Pareto dominated and that the game has no strong equilibrium. Prove that there are exactly four imputations and that the α-core coïncides with the set of imputations.

2) Each player must pick one among the three players, possibly himself. Hence

$X_i = \{1, 2, 3\}$

The payoffs of the game are deduced from those listed below by the symmetrical character of our game :

(x_1, x_2, x_3)	$= (1, 2, 3)$	payoffs :	$(0, 0, 0)$
	$= (1, 2, 1)$	"	$(0, 0, -1)$
	$= (1, 3, 1)$	"	$(0, 0, 0)$
	$= (1, 1, 1)$	"	$(3, 1, 1)$
	$= (1, 3, 2)$	"	$(0, 2, 2)$
	$= (2, 3, 1)$	"	$(2, 2, 2)$
	$= (2, 3, 2)$	"	$(-1, 3, 3)$

Prove that our game has no Nash and no strong equilibrium outcome.

Prove that there are exactly 5 imputations taking either one of the following forms :

(1, 1, 1) the player's vote is unanimous

(2, 3, 1) each player receives exactly one vote.

Prove that the α - core is made up of the two outcomes like (2, 3, 1).

Problem 1. The Nakamura theorem. (Nakamura [1979]).

Given a finite society N, a proper simple game is a subset W of non-empty coalitions of N such that :

$$\begin{cases} [T \in W,\ T \subset T'] \Rightarrow T' \in W \\ T \in W \Rightarrow T^c \notin W \end{cases}$$

Given a finite set A of candidates, a voting rule of N on A is a mapping π from A^N into A (each player casts the name of one candidate). We say that π is derived from W if the following property holds :

$$\forall x \in A^N \quad \forall a \in A \ [\exists T \in W \quad \forall i \in T \quad x_i = a] \Rightarrow [\pi(x) = a]$$

We denote by L(A) the set of linear orderings on A and by $u \in L(A)^N$ a preference profile over A : for all $i \in N$, u_i represents the utility of player i for the elements of A (indifferences are ruled out).

Given W and a voting rule π derived from W we associate to each particular preference profile u the game G(u) :

strategy set $X_i = A$ utility of i : $u_i \circ \pi$, all $i \in N$

1) Denote by C(u) the following, possibly empty, subset of A :

$$a \in C(u) \Leftrightarrow \forall b \in A \ \{i \in N / u_i(a) < u_i(b)\} \notin W$$

Prove that C(u) equals the image by π of the α-core of G(u) :

$$C(u) = \pi[C_\alpha(G(u))]$$

If the simple game W is strong :

$$T \notin W \Rightarrow T^c \in W$$

prove that :

$$C_\alpha(G(u)) = \pi^{-1}(C(u))$$

2) Suppose that C(u) is empty for some profile u. Prove the existence of p coaltions T_k , $k = 1,\ldots,p$, not necessarily distinct, such that :

$$\begin{cases} T_k \in W & \text{all } k = 1,\ldots,p \\ \bigcap_{k=1}^{p} T_k = \phi \end{cases}$$

3) Assume $|N| = n \geq p$ and prove the following equivalence (Nakamura's theorem) :

$$[\forall u \in L(A)^N \quad C(u) \neq \phi] \Leftrightarrow [|A| < \nu(W)]$$

where $\nu(W)$, the Nakamura' s number of W, is the minimal cardinality of a subset (T_α) of W such that :

$$\bigcap_\alpha T_\alpha = \phi$$

Interpretation ?

Problem 2. *Cooperative voting by veto* (*Moulin - Peleg* [*1980*]).

We fix a society N and the set A of candidates and we assume :

$$|N| = |A| - 1 \qquad (10)$$

A mapping π from A^N into A is said to be <u>of the voting by veto type</u> if it satisfies :

$$\forall x \in A^N \quad \forall i \in N \quad \pi(x) \neq x_i$$

From (10), such voting rules exist.

1) Given a preference profile $u \in L(A)^N$ (see notations of Problem 1) we denote by $CV(u)$ the following subset of A :

$$a \in CV(u) \Leftrightarrow \forall T \subseteq N \quad |T| + |P(T,a,u)| \leqslant p - 1$$

where $P(T,a,u) = \{b \in A / \forall i \in T \quad u_i(a) < u_i(b)\}$ is the set of candidates unanimously preferred to a by the members of T.

Prove that $CV(u)$ is always non-empty.

<u>Hint</u> : Consider the sequence $a_1, \dots, a_n$ defined inductively by :

a_1 = the worst candidate of u_1 among A

a_{t+1} = the worst candidate of u_{t+1} among $A \setminus \{a_1, \dots, a_t\}$

Next prove that $a = A \backslash \{a_1, \ldots, a_n\}$ belongs to $C\,V(a)$.

2) We denote by $G(u)$ the game with strategy sets $X_i = A$ and utility $u_i \circ \pi$, all $i \in N$.

Prove that the α - core of $G(u)$ is the inverse image of $C\,V(u)$ by π :

$$C_\alpha\,(G(u)) = \pi^{-1}\,(C\,V(u))$$

3) Prove that $C\,V(u)$ is also the image by π of the strong equilibrium set of $G(u)$:

$$\pi(\,SE(\,G(u))) = C\,V(u)$$

Hint : Pick $a \in C\,V(u)$ and denote

$$Q_i = \{b \in A \,/\, u_i(b) < u_i(a)\}$$

The Q_i are n subsets of the set $A \backslash \{a\}$ with cardinality n and satisfy moreover :

$$\forall\, T \subset N \quad \left| \bigcup_{i \in T} Q_i \right| \geq |T|$$

Thus by the marriage Lemma, there exists a labeling $a_1, a_2, \ldots, a_n$ of $A \backslash \{a\}$ such that $a_i \in Q_i$ for all $i \in N$. Next prove that $(a_1, \ldots, a_n) \in SE(\,G(u))$.

Problem 3. Objections and counter objections in three players games. (Laffond-Moulin [1980]).

Let $(X_1, X_2, X_3 ; u_1, u_2, u_3)$ be a three players game with finite strategy sets X_i. Given an outcome $x \in X_{\{1,2,3\}}$ we denote by $O_{12}(x)$ the set of objections by coalition $\{1,2\}$ against x :

$$[(y_1,y_2) \in O_{12}(x)] \Leftrightarrow [\inf_{y_3} u_i(y_1,y_2,y_3) > u_i(x_1, x_2, x_3) \quad \text{for } i = 1,2]$$

A counter objection by coalition $\{2,3\}$ against objection $(y_1, y_2) \in O_{12}(x)$ is a pair $(z_2, z_3) \in X_2 \times X_3$ such that :

$$\begin{cases} \inf_{z_1} u_2(z_1, z_2, z_3) > \inf_{y_3} u_2(y_1, y_2, y_3) \\ \sup_{z_1} u_1(z_1, z_2, z_3) < u_1(x_1, x_2, x_3) \end{cases} \qquad (11)$$

Player 3 forces the cooperation of 2 : if you stick to objection (y_1, y_2), I will keep your utility level not above $\inf_{y_3} u_2(y_1, y_2, y_3)$, whereas by agreeing to join me in (z_2, z_3) you are guaranteed of the utility level $\inf_{z_1} u_2(z_1, z_2, z_3)$. Finally the bottom inequality in (11) deters player 1 from entering objection (y_1, y_2).

The <u>deterrence set</u> of G is made up of all imputations such that to any objection by $\{i,j\}$ there exists a counter objection by $\{i, k\}$ and / or by $\{j, k\}$.

Prove that the deterrence set of G is non empty.

Hint : Assume first $I(G) = X_{\{123\}}$. Say that an objection $(y_1, y_2) \in O_{12}(x)$ by coalition $\{12\}$ against x is maximal if :

- it has no counter objection
- for all $(z_1, z_2) \in O_{12}(x)$:

$$\{\inf_{z_3} u_i(z_1, z_2, z_3) \geq \inf_{y_3} u_i(y_1, y_2, y_3) \text{ for } i = 1,2\}$$
$$\Rightarrow \{\text{these inequalities are equalities}\}$$

Suppose next that the deterrence set of G is empty and construct inductively a sequence

$$x^o, (S^1, x^1), (S^2, x^2), \ldots, (S^t, x^t), \ldots$$

where :

x^o is arbitrary in $X_{\{123\}}$

$x^t_{S_t}$ is a maximal objection of coalition S^t against x^{t-1} with no counterobjection (for all $t \geq 1$)

Then prove that any two outcomes x^t differ and contradict the finiteness of $X_{\{123\}}$.

III. REPEATED GAMES

Definition 4.

Given a N - normal form game $G = (X_i, u_i ; i \in N)$, the β - core of G is the subset, denoted $C_\beta(G)$, of those outcomes x^* such that :

for all coalition T, $T \subset N$, there exists a joint strategy $x_{T^c} \in X_{T^c}$ such that for all joint strategy $x_T \in X_T$:

$$\text{No} \begin{cases} u_i(x_T, x_{T^c}) \geq u_i(x^*) & \text{for all } i \in T \\ u_i(x_T, x_{T^c}) > u_i(x^*) & \text{for at least one } i \in T \end{cases}$$

The stability property of an outcome in the β - core is stronger than that of the α - core : a deviating coalition T can be countered by the complement coalition T^c even if the members of T keep secret their joint strategy x_T ; of course the fact that T is plotting a switch from the agreed upon strategy $x^*_{T^c}$ is public, otherwise the players in T^c are not aware of the betrayal and can not counter it....

Comparing Definition 3, 4 and Definition 1 Chapter V yields the inclusions :

$$S\,E\,(G) \subset C_\beta(G) \subset C_\alpha(G)$$

To interpret further Definition 4 we imagine that our game G is repeatedly played over time. At time t ,

$t = 1,2,...,$ each player i, aware of the past moves $x^1, ..., x^{t-1}$, selects a strategy x_i^t. The choices of the various players at time t are mixed or correlated strategies and are made independently or after some communication process; the actual drawings of these lotteries in period t are revealed publicly before the choice of period $t + 1$ takes place. Finally the payoff to each player i is the Cesaro mean of the instantaneous payoffs, that is to say :

$$\lim_{t \to +\infty} \frac{1}{t} \sum_{s=1}^{t} u_i(x^s) \tag{12}$$

In that framework, Aumann [1978] and Rubinstein [1979] prove that all outcomes in the mixed β-core of G are achieved by a strong equilibrium of the repeated game $G(\infty)$, and conversely the strong equilibrium payoffs of $G(\infty)$ cover the convex hull of the mixed β-core payoffs of G. Thus repetition of our original game formalizes an essential feature of cooperation by threats : deviations that are profitable in the short-run can prove harmful in the long run after the announced threat has been carried out. To say so, it is necessary that the long run payoffs infinitely outweight the short run payoffs, as is implicit in our interpretation of deterring scenarios.

The Cesaro-mean limit (12) achieves rigorously this condition.

Instead of sketching the (difficult) proof of the above mentioned results, we give intuition of them in the easier context where the overall payoff of the repeated game is the discounted sum of the instantaneous ones.

Repeated game with discounted payoff.

We start with an initial game $G = (X_i, u_i ; i \in N)$ where the u_i are uniformly bounded on X_N (e.g. X_i are compact and u_i continuous, all $i \in N$). We imagine that G is played once at $t = 1$, and let x^1 be the corresponding outcome. Next a random device commands to stop with probability $(1 - \delta)$ in which case the overall utility to player i is $u_i(x^1)$, all $i \in N$, or (with probability δ) to continue, in which case G is played again at time 2. And so on... After each play of G, it is decided with probability $(1- \delta)$ that the corresponding outcome is the final outcome or with probability δ, that the entire history of plays is neglected and another play of G occurs.

Another, formally equivalent, story goes by saying that G is played infinitely many times anyway, and the overall payoff resulting from the sequence of plays $x^1, x^2, \ldots, x^t$, to player i is :

$$(1-\delta) \ \{u_i(x^1) + \delta u_i(x^2) + \ldots + \delta^{t-1} u_i(x^t) + \ldots\} \text{ all } i \in N \quad (13)$$

When δ goes to 1 it is well known that the above summation approaches the Cesaro-mean limit (12) if the latter exists.

We set now $\beta_i = \inf_{x_{\hat{i}}} \sup_{x_i} u_i$ to be the maximal secure utility level of player i, i.e. the minimal payoff to player i when he can observe beforehand the strategies of all other players. Now we pick an imputation x^* such that

$$\beta_i \leq u_i(x^*) \qquad \text{all } i \in N$$

We interpret x^* as an agreement that the players wish to enforce by discounted repetition. For that purpose they will seek for a N E outcome σ^* of the repeated game which yields exactly payoff $u_i(x^*)$ to player i, all $i \in N$. Notice that σ_i^*, a strategy of player i in the repeated game, might be a complicated object that specifies for each past history $x^1, \ldots, x^{t-1}$ the strategy x_i^t that player i uses at time t (all $t \in \mathbb{N}$). Actually to enforce $u_N(x^*)$ as a N E payoff vector a very simple outcome σ^* is in order. At σ_i^* player i uses the agreed upon pure strategy x_i^* as long as no player j deviates from x_j^*. As soon as a deviating player j is detected, strategy σ_i^* commands to punish him for ever. Since we look at Nash equilibrium only, coalitional deviations from σ^* do not matter.

To describe formally our N E outcome, we pick for all $i \in N$ a strategy $N \setminus \{j\}$ - uple $\tilde{x}^j_{\hat{j}}$ such that :

$$\sup_{x_j} u_j(x_j, \tilde{x}^j_{\hat{j}}) = \beta_i$$

For all $i \in N$ a strategy σ_i^* of player i in the repeated game is choosen such that :

$x_i^1 = x_i^\star$: at time 1, play $x_i^\star$

if $x^1 = x^2 = \ldots = x^{t-1} = x^\star$ then $x_i^t = x_i^\star$: at time t play the agreed upon outcome if everybody did the same so far

if $x^1 = x^2 = \ldots = x^{t-2} = x^\star \neq x^{t-1}$ then pick a j such that $x_j^{t-1} \neq x_j^\star$ next play $\tilde{x}_i^j = x_i^t = x_i^{t+1} = \ldots$

Under what conditions is it the case that the behaviour depicted by $\sigma^\star$ deters any isolated individual from deviating ? It has to be that the short-run profit to player i of a deviation at time t (that can not be countered before the next play) is overweighted by the long-run loss suffered from the punishing behaviour by the players in $N \setminus \{i\}$ (who, by assumption stick to strategy $\sigma^\star_{N\setminus\{i\}}$) at time $t+1$, $t+2$, ... Setting :

$$u_i^\star(x_{\hat{i}}^\star) = \sup_{x_i} u_i(x_i, x_{\hat{i}}^\star)$$

and comparing the discounted value at time t of the short-run profit $u_i^\star(x_{\hat{i}}^\star) - u_i(x^\star)$ and the long-run losses, the self-enforcement property is then :

$$u_i^\star(x_{\hat{i}}^\star) - u_i(x^\star) \leq \delta(u_i(x^\star) - \beta_i) + \delta^2(u_i(x^\star) - \beta_i) + \ldots$$

which is equivalent to :

$$1 - \delta \leq \frac{u_i(x^\star) - \beta_i}{u_i^\star(x_{\hat{i}}^\star) - \beta_i} \quad \text{all } i \in N \tag{14}$$

Thus if δ is close enough to 1 conditions (14) are satisfied and σ^* is therefore a NE outcome of the repeated game (with discounted payoff (13) and/or with Cesaro mean limit payoff (12)).

<u>Lemma 4</u>.

If system (14) holds, repetition of G with discounted payoff (13) has a Nash equilibrium outcome of which the corresponding sequence of play is

$$x^* = x^1 = x^2 = \ldots = x^t = \ldots$$

and therefore the NE payoff to player i is $u_i(x^*)$

Notice that we have not described the whole set of NE outcomes of the repeated version of G. It turns out that the set of NE payoff vectors might not vary continuously with respect to δ. See Aumann [1978].

Our last example illustrates another simple approach to repetition, where each player takes only into account the last observed move of the other player.

Example 4. Repetition of the Prisonners' Dilemma.

(Aumann [1978]).

In the Prisonners' dilemma (Example 1, Chapter I) repeated over time, we assume that each player uses a stationary strategy with length 1 memory. Thus player i' strategy is a triple $(x_i ; y_i , z_i)$ where x_i, y_i, z_i all belong to $\{A, P\}$, to be interpreted as

- player i plays $x_i = x_i^1$ in the first occurrence of the game (time t = 1)
- at time $t \geqslant 2$, he plays y_i if player j was peaceful at time $t-1$, and he plays agressive if player j was agressive.

$$x_i^t = y_i \quad \text{if} \quad x_j^{t-1} = P$$

$$x_i^t = z_i \quad \text{if} \quad x_j^{t-1} = A$$

A typical strategy is "tit-for-tat" namely (P ; P, A).

After each player has choosen a strategy $(x_i ; y_i , z_i)$ a unique sequence $x^1, \ldots, x^t, \ldots$ results and the average payoff (12) is the final payoff. Hence the following (8 x 8) bimatrix game.

	a	*b*	*c*	*d*	*e*	*f*	*g*	*h*
a	3, 3	0, 4	3, 3	3, 3	0, 4	0, 4	0, 4	3, 3
b	4, 0	1, 1	1, 1	1, 1	4, 0	4, 0	1, 1	4, 0
c	3, 3	1, 1	3, 3	2, 2	2, 2	2, 2	1, 1	3, 3
d	3, 3	1, 1	2, 2	1, 1	2, 2	2, 2	1, 1	3, 3
e	4, 0	0, 4	2, 2	2, 2	2, 2	0, 4	0, 4	4, 0
f	4, 0	0, 4	2, 2	2, 2	4, 0	2, 2	0, 4	4, 0
g	4, 0	1, 1	1, 1	1, 1	4, 0	4, 0	1, 1	4, 0
h	3, 3	0, 4	3, 3	3, 3	0, 4	0, 4	0, 4	3, 3

(15)

a = P
b = A
c = P ; P , A
d = A ; P , A
e = P ; A , P
f = A ; A , P
g = P ; A , A
h = A ; P , P

P is a short hand for $(P ; P, P)$ identified with the "pure" peaceful strategy. Similarly $A = (A ; A, A)$ stands for the pure agressive strategy. Hence the North-West 2×2 submatrix of (15) is the original Prisonners' dilemma. Notice that for computational convenience we have taken slightly different numerical values from Example 1, Chapter I (see Exercize 4 below).

The payoff vector is 2×2 when the corresponding sequence (x^t) cycles. It can do so in three different ways :

- (P, A) (A, P) (P, A),...,for instance (P ; P, A) versus (A ; P, A)

or

- (P, P) (P, A) (A, A) (A, P) (P, P),..., for instance $(x_1 ; P, A)$ versus $(x_2 ; A, P)$ for all x_1, x_2

or

- (P, P) (A, A) (P, P),...,for instance (P ; A, P) versus (P ; A, P)

In all other cases the sequence (x^t) is stationary after finitely many steps and the corresponding payoff is "pure".

In the above 8×8 game, two strategies of each player can be deleted immediately namely (P ; A , A) as equivalent to A and (A ; P, P) as equivalent to P.

This leaves a 6×6 game with <u>two</u> Nash equilibrium outcomes, namely (A, A), the non-cooperative equilibrium of the original game and (P ; P, A) (P ; P, A) , a new N E outcome, where both players play "tit-for-tat". Interpretation of this specific self-enforcing agreement of the repeated game exactly parallels that of the deterring scenario of Example 1.

Note that in game (15) strategy P (peaceful) is no longer dominated by A (agressive). However after successive elimination of dominated strategies, P is eventually deleted. Actually the Pareto optimal NE outcome (i.e. tit-for-tat by both players) turns out to be the sophisticated equilibrium outcome of (15).

Exercise 4.

Prove that the game (15) is dominance-solvable with sophisticated equilibrium payoff (3,3). Is this result affected if the original game has the Prisonners' dilemma configuration but different cardinal utilities, thus giving distinct average payoffs to the three types of limit cycles ?

IV. CLASSIFICATION OF TWO-PERSON GAMES

In this section we restrict ourselves to two-person games. There the α - core is simply the imputation set with the β - core and the s - core (Definition 5) as two subsets, one of which at least is non empty (Theorem 1).

From now on we suppose given a <u>two</u> person game $G = (X_1, X_2, u_1, u_2)$ with <u>finite</u> strategy sets X_1, X_2 (although most of our results below are preserved for compact X_i and continuous u_i, $i = 1,2$.

<u>Lemma 5.</u>

The α - core of G equals the set of its imputations

$$C_\alpha (G) = I(G)$$

The β - core of G is given by :

$$x^* \in C_\beta(G) \Leftrightarrow \begin{cases} x^* \text{ Pareto optimal} \\ \inf_{y_i} \sup_{y_j} u_i(y_i, y_j) \leqslant u_i(x) \text{ for } i = 1,2 \end{cases}$$

The proof follows immediately Definition 3 and 4 and is therefore omitted.

Remark that the β - core of G can be empty : for instance choose for G a two person zero sum game without a value.

<u>Example 5.</u> <u>A perturbed coordination game.</u>

Both players pick an integer in $\{1,2,\ldots,10\}$. If x_1, x_2 are choosen and $x_1 + x_2 = 10$ then player i' payoff is x_i. Otherwise the payoff vector is (4, 0) if $x_1 + x_2$ is even and (0, 4) if $x_1 + x_2$ is odd.

This game is a mixture of a coordination game and "matching pennies".

The secure utility level sup inf u_i is zero, i.e. the minimal utility level. Hence the imputations are the Pareto optimal outcomes , i.e. the outcomes (x_1, x_2) such that $x_1 + x_2 = 10$.

However $\inf_{x_j} \sup_{x_i} u_i = 4$, $i = 1, 2$ so that by Lemma 4 the β- core contains only three outcomes

(4, 6) (5, 5) (6, 4)

To stabilize by deterring threats an imputation such as (2,8) it is necessary that both players accept to loose any privacy of their strategic choices. On the contrary an imputation in the β - core requires a less binding informational constraint : a signal to inform any player that his opponent has deviated in some unspecified way.

On the other hand in many two-person games the α - core and the β - core coïncide. This is true in a mixed extension, since the zero-sum games $(M_1, M_2, \bar{u}_1)$ and $(M_1, M_2, -\bar{u}_2)$ both have a value.

We consider next the subset of those imputations that can be stabilized by a pair of warnings i.e. a pair of threats that coïncide with the best reply function of each player.

Definition 5.

The s-core of G is the set of those imputations x^* such that there exists a deterring scenario (x^*, ξ_1, ξ_2) where ξ_i, $i = 1, 2$, is a warning :

$$\forall x_j \neq x_j^* \quad \begin{cases} u_j(x_j, \xi_i(x_j)) \leq u_j(x^*) \\ u_i(x_j, \xi_i(x_j)) = \sup\limits_{x_i \in X_i} u_i(x_j, x_i) \end{cases}$$

It is denoted $C_s(G)$

Lemma 6.

Suppose that u_i is one-to-one, $i = 1, 2$, and recall that S_i denotes i-Stakelberg payoff

$$S_i = \sup \{ u_i(x_i, x_j) \ / \ (x_i, x_j) \in BR_j \} \qquad i = 1, 2$$

where BR_j is the best reply correspondence of player j (see Chapter II, Section 4).

Then the s-core of G is made up of those Pareto optimal outcomes x such that :

$$S_i \leq u_i(x) \qquad i = 1, 2 \tag{16}$$

Proof.

Suppose x^* is a Pareto optimum satisfying (16) and denote for all $x_i \neq x_i^*$ the best reply strategy x_j of player j by $x_j = \xi_j(x_i)$. By definition of S_i we have then

$$u_i(x_i, \xi_j(x_i)) \leq S_i \leq u_i(x^*), \text{ all } i = 1,2, \text{ all } x_i \neq x_i^*.$$

Therefore (x^*, ξ_1, ξ_2) is a deterring scenario, hence x^* is individually rational (by Lemma 1 ; we could also invoke the always true inequality $\sup_{x_i} \inf_{x_j} u_i \leq S_i$). Finally x^* is Pareto optimal by assumption.

Conversely let x^* be an outcome in the $\mathfrak{s}$ - core of G. There exists a deterring scenario (x^*, ξ_1, ξ_2) where $\xi_i(x_j)$ equals the (unique by the one-to-one assumption) best reply strategy of player i to x_j. We prove that x^* satisfies (16). First pick $x_i \neq x_i^*$. Then by the deterring property of ξ_j we have :

$$u_i(x_i, \xi_j(x_i)) \leq u_i(x^*)$$

It remains to prove

$$x_j = \xi_j(x_i^*) \Rightarrow u_i(x_i^*, x_j) \leq u_i(x^*)$$

Suppose, on the contrary, $u_i(x^*) < u_i(x_i^*, x_j)$. Because x^* is Pareto optimal, we get :

$$\sup_{y_j} u_j(x_i^*, y_j) = u_j(x_i^*, x_j) < u_j(x^*)$$

a contradiction. ■

Remark 2.

The one-to-one assumption can be dropped in the statement of Lemma 6 if we simply replace S_i by

$$S'_i = \sup_{x_i} \inf_{x_j \in BR_j(x_i)} u_i(x_i, x_j)$$

The parallel proof is left as an exercize to the reader.

A warning is a selection of the best reply correspondence, and therefore a highly credible threat : a typical example is the tit-for-tat threat in Prisonners' dilemma ((1) in Example 1). Conversely a stabilizing threat which differs from a warning involves some pure deterrence feature such as : if you deviate in this way, I shall sacrify my short-run interest to the long-run duty of cooperative stability. Schelling [1971] gives the striking example of this cast of hindouist monks who weaponless protected traveling valuables by the mere threat to commit suicide if the valuables were ever stolen. A tradition of moral rigidity was enough to make the threat convincing.

Lemma 5 logically relates the feasibility of stability by warnings with the struggles for the leadership. (Chapter III, Section 1). It states that the struggle for the leadership occurs in G if and only if its s - core is empty. Thus in a game with an empty s - core, a commitment tactic threatens the stability of any outcome and the partial loss of rationality that makes a commitment believed is identical to the "doomsday machine" trick that makes a non-warning threat convincing.

Example 6. A quantity-setting duopoly.

We consider a duopoly game similar to Example 5, Chapter III. Two firms choose the quantity x_i of a given good that they supply on a market where the price $p_o - (x_1 + x_2)$ emerges (for $x_1 + x_2 \leqslant p_o$). Both firms have the same cost function, with a marginal cost linear with respect to the quantity produced. Increasing or decreasing returns to scale can occur.

$$\begin{cases} X_1 = X_2 = [0, \frac{1}{2} p_o] \\ u_i^{\varepsilon}(x_1, x_2) = [p_o - \bar{x}] x_i - (c + \varepsilon x_i) x_i \qquad i = 1,2 \end{cases}$$

where $\bar{x} = x_1 + x_2$ and p_o, c, ε are fixed parameters such that

$$0 < \frac{1}{2} p_o < c < p_o \qquad \varepsilon \text{ small w.r.t. } p_o - c$$

Case 1. Constant returns to scale : $\varepsilon = 0$

Observe that

$$u_1^o + u_2^o = [p_o - \bar{x}] \bar{x} - c \bar{x} = -\bar{x}^2 + (p_o - c) \bar{x}$$

therefore the outcomes x such that $\bar{x} = \frac{1}{2}(p_o - c)$ are Pareto optimal. Conversely, we let the reader check that all Pareto optimum are such that $\bar{x} = \frac{1}{2}(p_o - c)$ (by making the determinant of $\frac{\partial(u_1, u_2)}{\partial(x_1, x_2)}$ equal zero). The maximal joint profit is worth :

$$\max(u_1^o, u_2^o) = \frac{1}{4}(p-c)^2$$

The secure utility level is 0 (since $\sup_{x_1} \inf_{x_2} u_1^o = = \sup_{x_1} u_1^o(x_1, \frac{1}{2}p_o))$, thus the imputations yield arbitrary divisions of the maximal joint profit (provided that no one ends up with a negative profit).

To compute the Stackelberg - payoffs, remark that player 1's best reply is :

$$BR_1(x_2) = \frac{1}{2}(p-c) - \frac{1}{2}x_2$$

Thus :

$$S_2 = \sup_{0 \leq x_2 \leq \frac{1}{2}p_o} u_2^o(BR_1(x_2), x_2) =$$

$$= \sup_{0 \leq x_2 \leq \frac{1}{2}p_o} \frac{1}{2}[(p-c)x_2 - x_2^2] \Rightarrow S_2 = \frac{1}{8}(p-c)^2$$

Symmetrically we get $S_1 = \frac{1}{8}(p-c)^2$ hence the profit vector (S_1, S_2) is that of an imputation : by Lemma 5 the s- core is a singleton, namely $(x_1, x_2) = (\frac{1}{4}(p_o-c), \frac{1}{4}p_o-c))$ allocating half of the maximal joint profit to each player. In this symmetrical game the s-core is the fair cooperative outcome enforced by the natural threats BR_1, BR_2. Any unfair division of the maximal joint profit must be enforced by purely deterring, less convincing threats.

Case 2. Increasing or decreasing returns to scale.

A similar computation, using first order approximations (since ε is small) gives :

$$S_1(\varepsilon) = S_2(\varepsilon) \doteq \frac{1}{8}(p-c)^2(1+3\varepsilon)$$

Suppose first $\varepsilon > 0$. By definition of u_i^{ε}, we get

$$u_i^{\varepsilon}(x) < u_i^{o}(x), \quad \text{for all} \quad x_1 > 0 \quad x_2 > 0$$

This implies that the imputation payoff vector $(S_1(0), S_2(0))$ of $G(0)$ is no longer feasible in $G(\varepsilon)$, $\varepsilon > 0$. A fortiori $(S_1(\varepsilon), S_2(\varepsilon))$ is not feasible and the s-core of $G(\varepsilon)$ is empty.

A similar argument (invoking $S_i(\varepsilon) < S_i(0)$ and $u_i^{\varepsilon} > u_i^{o}$ for $\varepsilon < 0$) shows that the s - core of $G(\varepsilon)$ is <u>non</u> empty for increasing returns to scale ($\varepsilon < 0$).

Figure 2 illustrates the contrast between increasing and decreasing returns to scale.

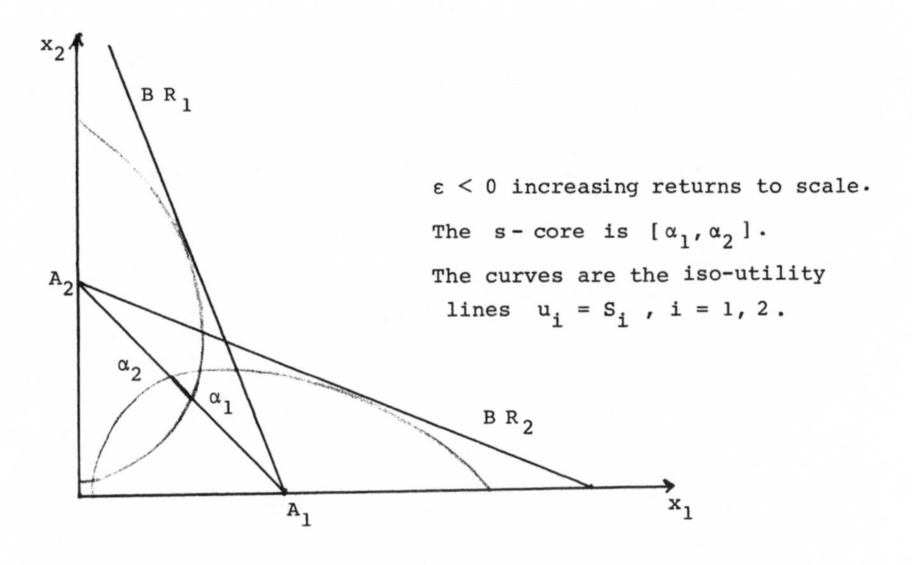
x_2
BR_1
A_2
α_2
α_1
BR_2
A_1
x_1
$\varepsilon < 0$ increasing returns to scale.
The s-core is $[\alpha_1, \alpha_2]$.
The curves are the iso-utility lines $u_i = S_i$, $i = 1, 2$.

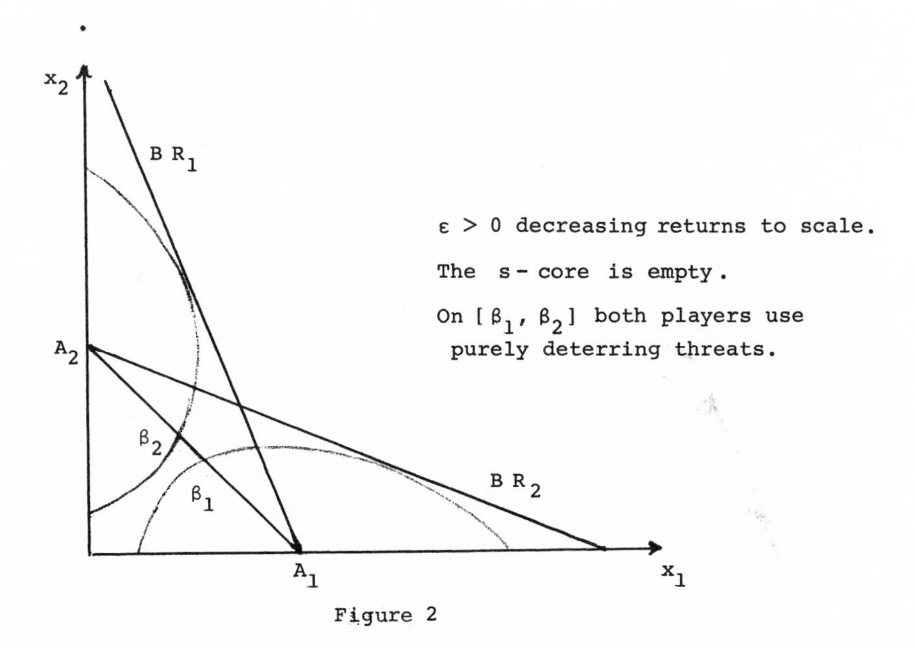
x_2
BR_1
A_2
β_2
β_1
BR_2
A_1
x_1
$\varepsilon > 0$ decreasing returns to scale.
The s-core is empty.
On $[\beta_1, \beta_2]$ both players use purely deterring threats.

Figure 2

Exercise 5.

In the perturbed coordination game (Example 5), show that the s-core is empty. Using the modified definition suggested in Remark 2, show that the s-core would be the fair imputation (5, 5).

Exercise 6.

Let $\| \ \|_i$ be an hilbert norm on $\mathbb{R}^2$ (i.e. derived from a positive definite symmetric 2 x 2 matrix), and $\alpha_i \in \mathbb{R}^2$, i = 1, 2, be fixed. Consider the game :

$$\begin{cases} X_1 = X_2 = \mathbb{R} \\ u_i(x) = -\| x - \alpha_i \| \qquad i = 1, 2 \end{cases}$$

Characterize geometrically and analytically the non-emptyness of the s-core.

From Lemma 6 the s-core of a given game G is empty if (and only if) the struggle for the leadership occurs in G. From Lemma 4 the β-core of G is empty if (and only if) the payoff vector $\gamma = (\inf_{y_2} \sup_{y_1} u_1, \inf_{y_1} \sup_{y_2} u_2)$ is not feasible (i.e. there exists no outcome x such that $\gamma \leq (u_1(x), u_2(x))$). By acting as a follower, player i is guaranteed of the utility level $\inf_{y_j} \sup_{y_i} u_i$ (since he knows of y_j before choosing y_i). Thus, emptyness of the β-core generates the "struggle for the followership" : whatever is the agreed upon outcome, at least one player would be better off by waiting until his opponent's strategy is irrevocably choosen and acting as a follower :

$$\forall x \in X_{\{12\}} \; \exists i \in \{1,2\} \quad u_i(x) < \inf_{y_j} \sup_{y_i} u_i$$

By our next result, this terminology is consistent : the struggle for the leader - and followership can not occur together.

Theorem 1.

The s-core and the β-core can not be simultaneously empty.

If the s-core and the β-core are non-empty, they intersect.

Corollary.

Two person games are partitionned into three classes :

i) $C_\beta = \phi$; $C_s \neq \phi$: struggle for the followership

ii) $C_s = \phi$; $C_\beta \neq \phi$: struggle for the leadership

iii) $C_\beta \cap C_s \neq \phi$: $C_\beta \cap C_s$ is the favourable area to cooperation by threats.

Proof.

Let x^i, $i = 1, 2$, be a Stackelberg equilibrium where i is leader let D_i be the subset of outcomes x such that :

$$S_i \leqslant u_i(x) \quad ; \quad \inf_{y_i} \sup_{y_j} u_j \leqslant u_j(x)$$

Since x_j^i is a best reply of player j to x_i^i, the set D^i contains x^i and is therefore non empty.

We suppose now that the s-core and the β-core are together empty and derive a contradiction. We pick a Pareto optimal outcome x in D_1 (for instance by maximizing $u_1 + u_2$ on D_1). Then we have :

$$C_s(G) = \phi \Rightarrow u_2(x) < S_2$$

$$C_\beta(G) = \phi \Rightarrow u_1(x) < \inf_{y_2} \sup_{y_1} u_1$$

Since x^2 is in D_2 we conclude that x^2 Pareto dominates x, a contradiction.

Suppose now that the s-core and the β-core are together non empty : we prove that their intersection is non-empty as well.

If $(S_1, S_2) \leqslant \gamma$ then $C_\beta \subset C_s$ hence the desired conclusion. Similarly if $\gamma \leqslant (S_1, S_2)$ we have $C_\beta \cap C_s = C_s$. It remains to consider the cases like :

$$\inf_{y_2} \sup_{y_1} u_1 < S_1 \quad ; \quad S_2 < \inf_{y_1} \sup_{y_2} u_2$$

Then any outcome in D^1, for instance x^1, belongs to $C_s \cap C_\beta$.

∎

Typical examples of games in class i) are zero-sum games without a value. Actually a game in class i) has no Nash equilibrium :

Exercise 7.

If the β-core of G is empty, then $NE(G)$ is empty as well and the zero sum games (X_1, X_2, u_1) and $(X_1, X_2, -u_2)$ have no value.

Examples of games in class ii) are the crossing game (Example 2 Chapter III) or the perturbed coordination game (Example 5).

Exercise 8.

If G belongs to class ii), we have :

$$\inf_{y_j} \sup_{y_i} u_i < S_i \quad \text{for } i = 1,2.$$

For a game in class iii) the intersection $C_s \cap C_\beta$ is the likely area of cooperation by threats. Typical examples are all games with a dominating strategy equilibrium or with a Pareto optimal payoff vector (S_1, S_2) and/or γ as the next two exercizes and the next problem illustrate.

Exercise 9.

If each player has a strictly dominating strategy :

$$u_i(x_i^\star, x_j) > u_i(x_i, x_j) \quad \text{all} \quad x_i \neq x_i^\star, \text{ all } x_j$$

then G belongs to class iii).

If each player has a dominating strategy, prove by an example that G can belong to class ii).

Exercise 10.

If (S_1, S_2) is the payoff vector of at least one imputation, then G belongs to class iii).

Problem 4. Quasi-inessential games.

Say that G is quasi-inessential if
$\gamma = (\inf_{y_2} \sup_{y_1} u_1, \inf_{y_1} \sup_{y_2} u_2)$ is the payoff of at least one Pareto optimal outcome.

1) Suppose that u_1, u_2 are one-to-one on $X_{\{12\}}$, and that G is q.i; Then the (unique) imputation x^* such that $(u_1(x^*), u_2(x^*)) = \gamma$ is an i - Stackelberg equilibrium for $i = 1, 2$. It is also the unique NE outcome of G.

2) Give an example of a q.i. game with no NE outcome.

3) Suppose that G is q.i.. Prove that G is inessential if and only if the zero-sum games (X_1, X_2, u_1) and $(X_1, X_2, -u_2)$ both have a value.

Give an example of a non inessential q.i. game where u_i is one-to-one, $i = 1, 2$.

Problem 5. Guaranteed deterring scenarios.

Moulin [1977].

Let (x^*, ξ_1, ξ_2) be a deterring scenario of G. We say that it is <u>guaranteed</u> if no player can suffer a loss by carrying out his threat :

$$\left.\begin{array}{l} u_1(x_1, \xi_2(x_1)) \leq u_1(x^*) \leq u_1(\xi_1(x_2), x_2) \\ u_2(\xi_1(x_2), x_2) \leq u_2(x^*) \leq u_2(x_1, \xi_2(x_1)) \end{array}\right\} \text{ all } x_1, x_2$$

The g- core of G is the subset, denoted $C_g(G)$, of those imputations x^* such that there exists at least one guaranteed deterring scenario (x^*, ξ_1, ξ_2).

1) Prove the following equivalence :

$$x \in C_g(G) \Leftrightarrow [\, x \text{ is Pareto optimal, and } u_i(x) \leq \inf_{y_j} \sup_{y_i} u_i \,]$$

2) Prove that the g-core is a - possibly empty - subset of the s-core :

$$C_g(G) \subset C_s(G)$$

3) If G belongs to class i), prove by examples that the g-core can be empty or non-empty.

Hint : consider a 2 x 2 game such as :

0 / 2	2 / 0
3 / 1	1 / 3

and describe the non binding agreements of this game (including mixed NE and correlated equilibria.

4) If G belongs to class iii) prove that either its g-core is empty or G is quasi-inessential (Problem 4). In the latter case prove that :

$$C_g = C_s = C_\beta = (u_1, u_2)^{-1}(\gamma)$$

IV. GAMES IN CHARACTERISTIC FORM : THE CORE

Example 7. A jazz band game. (Young [1979]).

A singer (S), a pianist (P), and a drummer (D) are offered \$ 1,000 to play together by a night club owner in Paris. The owner would alternatively pay \$ 800 the singer-piano duo, \$ 650 the piano drums duo, and \$ 300 the piano alone. Since he wants this piano he is not interested in any other duo or soloist. Moreover the singer-drums duo makes \$ 500 a night in one well located metro station, and the singer alone gets on average \$ 200 a night along the "terrasses de café". The drums alone can make no profit.

The three musisians get the maximal joint profit (1 000) by performing together in the night club : they get 650 + 200 if the singer splits from the piano drums duo, 300 + 500 if the piano is hired alone by the night-club, and 800 if the piano and singer reject the drums' partnership.

What division of the maximal joint profit \$ 1,000 can be said to be stable, given the opportunities for partial cooperation and/or selfish behaviour available to our three players ?

Definition 6.

Given a finite society N , a N - characteristic form

<u>game</u> is a mapping v from the set $\mathcal{P}(N)$ of non-empty coalitions of players into $\mathbb{R}$. The game (N, v) is said to be <u>super-additive</u> if v is such that for all disjoint $T, S \in \mathcal{P}(N)$ we have :

$$v(T) + v(S) \leq v(T \cup S) \tag{17}$$

Although super-additivity (condition (14)) is not technically necessary for most of our results below, it is needed to interpret the number $v(T)$ as the maximal joint profit resulting from cooperation within coalition T, independently of the complement coalition T^c. Inequality (17) then expresses that coalition $T \cup S$ can do at least as well as by letting T and S independently cooperate.

Super additivity of v clearly implies that for all partition $S_1, \ldots, S_K$ of N :

$$\sum_{k=1}^{K} v(S_k) \leq v(N) \tag{18}$$

Thus no partition $S_1, \ldots, S_K$ can yield a total joint profit above the profit $v(N)$ achieved by the grand coalition. Henceforth cooperation of all players is the only efficient cooperative outcome.

<u>Remark 1.</u>

Definition 1 supposes that the output of cooperation can be maesured in terms of a single numeraire, i.e. a transferable private good allowing side-payments among players. Note that we have on the contrary developped most of the theory of normal form games in the more general framework where individual utility functions are not taken to be transferable, or even comparable among players (see Chapters I, II, III and VI). Actually we restrict ourselves now to the transferable utility case for the sake of simplicity only. References for the theory of games in characteristic form without side-payments are to be found in Aumann [1976].

The next two definitions parallel that of the imputations and the α-core of normal form games.

Definition 8.

An imputation of the N-characteristic form game v is a vector $x = (x_i)_{i \in N}$ of $\mathbb{R}^N$ such that :

$$\sum_{i \in N} x_i = v(N) \quad ; \quad x_i \geqslant v(\{i\}) \quad \text{all } i \in N$$

We denote by $I(v)$ the set of imputations of game (N, v).

Applying iteratedly the super-additivity condition (17) yields :

$$\sum_{i \in N} v(\{i\}) \leqslant v(N)$$

Hence the set $I(v)$ of imputations is non empty for superadditive games.

Since $v(\{i\})$ is the secure profit level of which player i is guaranteed by himself, it plays the role of the secure utility level of normal form games. Lemma 2 above draws a rigorous analogy between the two notions of imputation.

If the game (N, v) is such that $\sum_{i \in N} v(\{i\}) = v(N)$, the set $I(v)$ shrinks to a singleton and the game (N,v) is said to be inessential (see the interpretation of Definition 5, Chapter I). From the superadditivity of v we deduce then that v is additive (by (18)) :

$$\text{for all } S \subseteq N \quad v(S) = \sum_{i \in S} v(\{i\})$$

(The proof of this claim is left as an exercize to the reader) For non inessential games, the core is a possibly empty subset of imputations stabilized by easy threats.

Definition 9.

The core of (N, v) is the subset of imputations $x \in I(G)$ such that :

$$\text{for all } S \subseteq N : \sum_{i \in S} x_i \geqslant v(S) \tag{19}$$

It is denoted $C(v)$.

Let our players bargain on the choice of a cooperative agreement. By super-additivity of v, the agreement requires the overall cooperation of the grand coalition N. The discussion bears on the division of the joint profit $v(N)$, i.e. the choice of a vector $x \in \mathbb{R}^N$ such that $\sum_{i \in N} x_i = v(N)$. Individual rationality, $v(\{i\}) \leqslant x_i$, all $i \in N$, is a minimal requirement to obtain the consent of player i, therefore the players bargain on the choice of a particular imputation x. Against x any coalition S could form and ask for a better allocation (e.g. some $y \in I(v)$ such that $y_i > x_i$, all $i \in S$), threatening to break down the overall cooperation (a very feasible threat since the unanimous consent of all players is needed to achieve the joint profit $v(N)$). To counter this threat suppose that the players in S^c react by refusing once and for all to cooperate with any member of S. Then coalition S is left afterall with a maximal joint profit $v(S)$ and condition (9) is the deterring property of that threat by S^c. Thus the core of (N, v) is the set of those divisions of $v(N)$ that are stabilized by the - natural - coalitional threats : no cooperation any more with any coalition claiming for a better share of $v(N)$.

Besides this interpretation of the core by deterring threats, we have an equivalent normative interpretation.

Lemma 7.

Let $x \in I(v)$ be an imputation of game (N, v). Then x belongs to the core $C(v)$ if and only if :

$$\text{for all } S \subset N \ : \ \sum_{i \in S} x_i \leqslant v(N) - v(S^c)$$

Proof.

Since $\sum_{i \in N} x_i = v(N)$, the above inequality is re-written as $v(S^c) \leqslant \sum_{i \in S^c} x_i$.

∎

Thus the core is made up of those imputations where no coalition S receives more than its marginal contribution $v(S \cup S^c) - v(S^c)$ to the profit of the grand coalition N.

Example 7.(continued)

A vector $x = (x_S, x_P, x_D)$ in our jazz band game (Example 7) belongs to the core if and only if :

$$\left.\begin{array}{l} x_S \geqslant 200 \ , \ x_P \geqslant 300 \ , \ x_D \geqslant 0 \\ x_S + x_P + x_D = 1\,000 \end{array}\right\} \text{imputation}$$

$$\left. x_S + x_P \geqslant 800, \ x_P + x_D \geqslant 650, \ x_S + x_D \geqslant 500 \right\} \text{core}$$

It is the convex hull of the following three imputations :

(350, 450, 200) (350, 500, 150) (300, 500, 200)

Thus all players' payoffs are determined up to \$ 50. A typical member of the core is the center of $C(v)$ namely :

$$x^{\star} = (333.3,\ 483.3,\ 183.3)$$

At $x^{\star}$, each 2 players coalition makes the same extra-profit $x_i + x_j - v(\{i,j\}) = 16.6$ \$. Imputation $x^{\star}$ is a fair compromise within $C(v)$.

Emptyness of the core does not imply that cooperation of the overall coalition N is impossible. It simply means that no imputation can be stabilized by the simple, natural threats described above. In this case cooperative stability requires a more complex behavioural scenario similar to the threat and counterthreats described in Section 2 (see in particular Problem 3). This leads to Aumann and Maschler [1964] bargaining set concept.

Emptyness of the core occurs when intermediate coalitions are too powerful. This is particularly clear in our next example.

<u>Example 8</u>. <u>Symmetric games</u>.

A symmetric game gives identical power to coalitions

with the same size. The characteristic form v is such that:

$$v(S) = v^{\star}(s) \quad \text{where} \quad s = |S| \text{ , all } S \subseteq N$$

Suppose, without loss of generality $v^{\star}(s) = 0$, and $N = \{1, 2, \ldots, n\}$. Then the imputation set of $v^{\star}$ is simply the following simplex of $\mathbb{R}^n$:

$$x \in I(v^{\star}) \Leftrightarrow \sum_{i=1}^{n} x_i = v^{\star}(n) \text{ , } x_i \geqslant 0 \text{ , } i = 1, \ldots, n.$$

The core $C(v^{\star})$ is a subset of $I(v^{\star})$ defined by finitely many (actually $2^n - n - 2$) linear inequalities (17). Hence it is a, possibly empty, convex polyhedron within the simplex $I(v^{\star})$. By symmetry of v, the core $C(v^{\star})$ is symmetrical, i.e. invariant under any permutation of the components $x_1, \ldots, x_n$. Together with convexity of $C(v^{\star})$, this implies that $C(v^{\star})$ is non empty if and only if it contains the center $x^{\star}$ of $I(v^{\star})$ namely $x_i^{\star} = \frac{1}{n}$, all $i = 1, \ldots, n$. Back to system (17) we conclude :

$$[C(v^{\star}) \neq \phi] \Leftrightarrow [\text{for all } s = 1, \ldots, n \quad \frac{1}{s} v^{\star}(s) \leqslant \frac{1}{n} v^{\star}(n)]$$

Thus the core $C(v^{\star})$ is non empty if and only if no intermediate coalition S is able to make a profit above its proportional share of the overall profit $v(N) = v^{\star}(n)$.

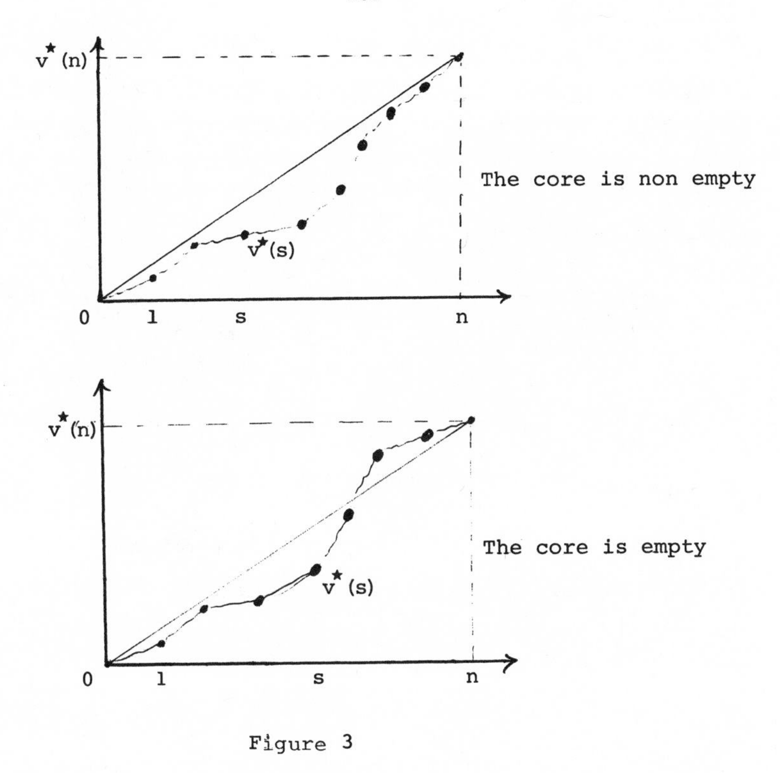

Figure 3

Back to the general case, we characterize now the non-emptyness of the core with the help of linear programming.

Definition 10.

Given N and a player i we denote by $\mathcal{P}(i)$ the set of non-empty coalition containing player i :

$$\{S \in \mathcal{P}(i)\} \Leftrightarrow \{S \in \mathcal{P}(N) \text{ and } i \in S\}$$

Now a balanced family of coalitions is a mapping δ from $\mathcal{P}(N)$ into $[0, 1]$ such that :

$$\text{for all } i \in N \quad \sum_{S \in \mathcal{P}(i)} \delta_S = 1$$

Finally we say that a game in characteristic form (N, v) is <u>balanced</u> if it satisfies :

$$\sum_{S \in \mathcal{P}(N)} \delta_S\, v(S) \leq v(N) \quad \text{for all balanced family of coalitions } \delta. \tag{20}$$

<u>Theorem 2</u>. (Bondavera, Scarf [1967]).

The game (N, v) has a non-empty core if and only if it is balanced.

<u>Proof</u>.

If x belongs to $C(v)$ and δ is a balanced family of coalitions we have :

$$x(S) \geq v(S) \Rightarrow \delta_S\, x(S) \geq \delta_S\, v(S) \quad , \text{all } S \subset N.$$

By summing up these inequalities we get :

$$\sum_{S \in \mathcal{P}(N)} \delta_S\, v(S) \leq \sum_{S \in \mathcal{P}(N)} \delta_S\, x(S) = \sum_{i \in N} \sum_{S \in \mathcal{P}(i)} \delta_S\, x_i = \sum_{i \in N} x_i = v(N)$$

Conversely if the core $C(v)$ is empty the hyperplane

$$\sum_{i \in N} x_i = v(N)$$

is disjoint from the convex (non empty) polyhedron :

$$\text{for all } S \subseteq N \qquad \sum_{i \in S} x_i \geqslant v(S)$$

By a standard separation argument, this implies the existence for all S of a non negative number δ_S such that :

$$\text{for all } x \in \mathbb{R}^N : \sum_{i \in N} x_i = \sum_{S \in P(N)} \delta_S \left(\sum_{i \in S} x_i \right)$$

and

$$\sum_{S \in P(N)} \delta_S \, v(S) > v(N)$$

The first of these properties is equivalent to saying that δ is a balanced family of coalitions. This concludes the proof of Theorem 2.

∎

As an application of Theorem 2, consider a three person game $|N| = 3$. To each partition of N is associated a trivial balanced family of coalitions :

$$\text{partition } N : \delta_N = 1 \quad \delta_S = 0 \quad \text{all } S \neq N$$

$$\text{partition } \{i\}\{jk\} : \quad \delta_{\{i\}} = 1 \quad \delta_{\{jk\}} = 1 \quad \delta_S = 0 \text{ otherwise.}$$

Applying (18) to these δ yields simply superadditivity of v :

$$v(i) + v(jk) \leqslant v(N) \quad \text{all } \{i,j,k\} = \{1,2,3\} \tag{21}$$

Next, to the coalition structure {12} {23} {31} is associated the balanced family of coalitions $\delta^{\star}$:

$$\delta^{\star}_{\{ij\}} = \frac{1}{2} \quad \text{all } i, j, i \neq j$$

$$\delta^{\star}_{S} = 0 \quad \text{otherwise}$$

Applying (20) to δ yields :

$$v(12) + v(23) + v(31) \leqslant 2\, v(123) \qquad (22)$$

Finally we let the reader check that all balanced families of coalitions on N are convex combinations of δ^N, $\delta^{\{i\}\ \{jk\}}$ and $\delta^{\star}$. Therefore system (21) (22) characterizes the non-emptyness of $C(v)$.

Corollary of Theorem 2.

A three player superadditive game has a non-empty core if and only if (22) holds true.

When the cardinality of N increases the balancedness of v becomes more and more difficult to compute. For instance a four players (superadditive) game has a non-empty core if and only if :

$$v(12) + v(23) + v(34) + v(41) \leqslant 2\, v(1234)$$

$$v(123) + v(234) + v(134) + v(124) \leqslant 3\, v(1234)$$

$$v(123) + v(234) + v(14) \leqslant 2\, v(1234)$$

and five similar inequalities derived by permuting the players.

Exercise 11 prove this assertion.

In Problem 8 below we give a typical application of Theorem 2 to economic models.

Problem 6. The core of simple games.

We say that game (N, v) is simple if :

for all $S \subseteq N$, $v(S) = 0$ or 1

In that case we denote by W the set of winning coalitions, namely :

$$S \in W \Leftrightarrow v(S) = 1$$

1) Prove that the simple game (N, v) is superadditive if and only if W is monotonic and proper :

$$S \in W, \; S \subseteq T \Rightarrow T \in W \quad \text{all } S, T$$

$$S \in W \Rightarrow S^c \notin W$$

2) Say that player $i^* \in N$ is a dictator of W if $\{i^*\}$ is a winning coalition :

$$\{i^*\} \in W$$

Assuming that (N, v) is superadditive, prove that W has a unique dictator if and only if it is inessential.

3) Suppose that W has no dictator and is non empty.

Say that player i^* is a veto player if $N \setminus \{i^*\}$ is <u>not</u> winning. Denote by $N_\star$ the, possibly empty, set of veto players.

Prove that the core $C(v)$ is non empty if and only if $N_\star$ is non empty. In this case prove :

$$\{x \in C(v)\} \Leftrightarrow \{\sum_{i \in N_\star} x_i = 1, \quad x_i = 0 \quad \text{all} \quad i \notin N_\star$$
$$, \quad x_i \geq 0 \quad \text{all} \quad i \in N\}$$

4) Let for all i, player i be endowed with a voting weight $q_i \geq 0$ and denote by $q_o \geq 0$ a quota such that :

$$q_o \leq \sum_{i \in N} q_i < 2\, q_o$$

Then define the weighted majority game W_q by :

$$S \in W_q \Leftrightarrow \sum_{i \in S} q_i \geq q_o$$

Prove that W_q is a monotonic and proper simple game. Prove that i^* is a dictator if and only if $q_o \leq q_{i^*}$ and i^* is a veto player if and only if :

$$\sum_{i \in N} q_i - q_o < q_{i^*}$$

5) Compare the above results with Problem 1, Chapter VI.

Problem 7. Convex games.

Let (N, v) be a given convex game i.e. such that the following property holds :

$$v(S) + v(T) \leq v(S \cap T) + v(S \cup T), \quad \text{all } S, T \subseteq N$$

1) Prove that :

$$v(N) - v(S^c) = \sup_{T \subseteq S^c} \{v(S \cup T) - v(T)\} \quad , \text{all } S \subset N$$

This property reinforces interpretations of Lemma 6.

2) For any ordering $1,2,\ldots,n$ of N, set :

$$x_i = v(1,2,\ldots,i) - v(1,2,\ldots,i-1) \qquad i = 2,\ldots,n$$

$$x_1 = v(1)$$

Prove that x belongs to the core of game (N, v).

3) Prove that $C(v)$ is the convex hull of the imputations x_σ obtained as in 2) when the ordering σ of N varies.

Problem 8. The core of an exchange economy.

Let for all $i \in N$, u_i be a utility function defined on the positive orthant $\mathbb{R}_+^p$. We interpret $z \in \mathbb{R}_+^p$ as a consumption vector and $u_i(z)$ as player i' utility for z.

Let for all $i \in N$, $\omega_i \in \mathbb{R}_+^p$ be player i' initial endowment.

The exchange economy game (with side-payments), associated with $(u_i, i \in N)$ is defined by :

for all $S \subseteq N$ $v(S) = \sup \{\sum_{i \in S} u_i(z_i) / \sum_{i \in S} z_i = \sum_{i \in S} \omega_i, z_i \in \mathbb{R}_+^p\}$

1) Suppose first $p = 1$ and $N = B \cup S$ where for $i \in B$ (buyer) $\omega_i = 0$ whereas for $i \in S$ (seller) $\omega_i = 1$.

Suppose that for all $i \in N$, u_i is given by :

$$\text{if } i \in B \quad \begin{cases} u_i(z) = 0 & \text{for } z < 1 \\ u_i(z) = b_i & \text{for } z \geqslant 1 \end{cases}$$

$$\text{if } j \in S \quad \begin{cases} u_j(z) = -s_j & \text{for } z < 1 \\ u_j(z) = 0 & \text{for } z \geqslant 1 \end{cases}$$

Finally we order B and N in such a way that :

$$b_{n_b} \leqslant \ldots \leqslant b_i \leqslant \ldots \leqslant b_1 \quad \text{for } i \in B = \{1,\ldots,n_b\}$$

$$s_1 \leqslant \ldots \leqslant s_j \leqslant \ldots \leqslant s_{m_s} \quad \text{for } j \in S = \{1,\ldots,m_s\}$$

Prove that $v(N) = \sum_{i=1}^{i^*} (b_i - s_i)$ where i^* is the greatest integer such that :

$$i^* \leqslant \inf(n_b, m_s), \quad s_{i^*} \leqslant b_{i^*}$$

Prove that if an imputation x belongs to C(v) there exists a price q such that :

$$\begin{cases} x_i = b_i - q & \text{for } i \in B, \quad i \leq i^* \\ x_i = 0 & \text{for } i \in B, \quad i > i^* \end{cases}$$

$$\begin{cases} x_j = q - s_j & \text{for } j \in S, \quad j \leq i^* \\ x_j = 0 & \text{for } j \in S, \quad j > i^* \end{cases}$$

Compute the interval $I = [q_{inf}, q_{max}]$ such that for $q \in I$ the imputation (23) belongs to $C(v)$.

Prove that all competitive equilibrium prices of this exchange economy belong to I.

2) We suppose now that p is finite and arbitrary, and that for all $i \in N$, u_i is a concave function over $\mathbb{R}_+^p$.

Prove that the game (N, v) is balanced and therefore has a non empty core. Prove that a competitive equilibrium allocation belongs to $C(v)$.

REFERENCES

ARON, R. 1962. Paix et guerre entre les nations. Paris : Calman-Levy Ed. .

AUMANN, R.J. 1976. Lecture on game thoery. Stanford : Stanford University, IMSSS..

AUMANN, R.J. 1978. Survey on repeated games. Stanford : Stanford University, IMSSS.. In Essays in Game Theory and Mathematical Economics in honor of Oskar Morgenstern, 1980, Bibliographisches Institut Mannheim.

AUMANN,R.J.and M.MASCHLER. 1964. The bargaining set for co-operative games. Advances in Game Theory. Annals of Math. Studies n° 52. Princeton N.J.: Princeton University Press.

BONDAVERA,O.N.. 1962. Teoriia idra v igre n lits (The theory of the Core in an n-person game). Vestnik Leningradskogo Universiteta Seriia Matematika, Mekhanika i Astronomii n° 13 : 141-142.

HOWARD, N. 1971. Paradoxes of rationality : theory of metagames and political behaviour. Cambridge (USA) : M.I.T.Press.

LUCE, R.D. and H. RAIFFA. 1957. Games and decisions. New York : John Wiley and Sons.

MOULIN, H. 1977. "Cours de théorie des jeux à deux joueurs" Cahier de Mathématiques de la Décision n° 7709. Paris : Université Dauphine.

MOULIN, H. and B. PELEG. 1980. "Stability and implementation of effectivity functions". Forthcoming in Journal of Mathematical Economics (1982).

MOULIN, H. 1981. "The strategy of social choice". Cahier du Laboratoire d'Econométrie de l'Ecole Polytechnique n° A229. Paris.

NAKAMURA, K. 1979. "The vetoers in a simple game with ordinal preferences". International Journal of Game Theory 8,1 : 55-61.

ROSENTHAL, R. 1972. "Cooperative games in effectiveness form". Journal of Economic Theory 5, 1.

ROTH, A. 1976. "Subsolutions and the supercore of cooperative games". Mathematics for Operation Research 1, 1.

RUBINSTEIN, A. 1979. "Equilibrium in supergames with the overtaking criterion". Journal of Economic Theory 21, 1 : 1-9.

SCARF, H. 1967. The core of an N-person game. Econometrica 35 : 50-69.

SCHELLING, T.C. 1971. The strategy of conflict . Cambridge (USA): Harvard University Press.

SEN, A. 1970. Collective Choice and Social Welfare. San Francisco : Holden Day.

SHERER, F. 1970. Industrial pricing. Chicago : Rand Mac Nally.

VICKREY, W.S. 1959. Self-policing properties of certain imputations sets. Annals of Math. Studies 40. Princeton : Princeton University Press.

YOUNG, H.P. 1979. "The market value of a game". IIASA working paper : Laxenburg.

I N D E X